Economics and Policy Issues in Climate Change

EDITED BY

William D. Nordhaus

Routledge
Taylor & Francis Group

LONDON AND NEW YORK

First published 1998 by Resources for the Future

2 Park Square, Milton Park, Abingdon, Oxfordshire OX14 4RN
711 Third Avenue, New York, NY 10017

Routledge is an imprint of the Taylor & Francis Group, an informa business

First issued in paperback 2018

Library of Congress Cataloging-in-Publication Data

Economics and policy issues in climate change / edited by William D. Nordhaus.
     p.   cm.
   Includes bibliographical references and index.
   ISBN 0–915707–95–0 (hardcover)
   1. Climatic changes—Economic aspects. 2. Climatic changes—Social aspects. 3. Climatic changes—Government policy. I. Nordhaus, William D.
QC981.8.C5E215   1998
363.738´742—dc21
                        98–27128
                        CIP

ISBN 978-0-915707-95-9 (hbk)
ISBN 978-1-138-37659-5 (pbk)

The paper in this book meets the guidelines for permanence and durability of the Committee on Production Guidelines for Book Longevity of the Council on Library Resources.

This book was typeset in Palatino by Betsy Kulamer, and its cover was designed by AURAS Design.

About
Resources for the Future

Resources for the Future is an independent nonprofit organization engaged in research and public education with issues concerning natural resources and the environment. Established in 1952, RFF provides knowledge that will help people to make better decisions about the conservation and use of such resources and the preservation of environmental quality.

RFF has pioneered the extension and sharpening of methods of economic analysis to meet the special needs of the fields of natural resources and the environment. Its scholars analyze issues involving forests, water, energy, minerals, transportation, sustainable development, and air pollution. They also examine, from the perspectives of economics and other disciplines, such topics as government regulation, risk, ecosystems and biodiversity, climate, Superfund, technology, and outer space.

Through the work of its scholars, RFF provides independent analysis to decisionmakers and the public. It publishes the findings of their research as books and in other formats, and communicates their work through conferences, seminars, workshops, and briefings. In serving as a source of new ideas and as an honest broker on matters of policy and governance, RFF is committed to elevating the public debate about natural resources and the environment.

Contents

Foreword

"Understanding the science, economics, and policy aspects of global warming has proved to be one of the most exciting and challenging tasks facing the natural and social sciences over the last decade." This somewhat understated opening sentence by William Nordhaus sets the stage for a rich, well-crafted book on the economic and policy dimensions of climate change. The problems are complex, and the stakes are high. The essays in this volume, and the comments on those essays, will provide valuable insights for analysts and the policymakers they advise for years to come.

The starting point for the volume Professor Nordhaus has assembled is a 1995 report on the economic and social dimensions of climate change by the Intergovernmental Panel on Climate Change (IPCC). This unusual institution was somewhat presciently created by the United Nations and the World Meteorological Organization in 1988, several years prior to the signing of the Framework Convention on Climate Change in 1992. The IPCC's charge is to provide high-quality, unbiased, and policy-relevant syntheses of knowledge concerning not just the science of climate change and its potential impacts, but also the socioeconomic consequences of climate change and greenhouse gas limitation policies. The IPCC's activities are carried out by international teams involving many of the world's best experts on topics almost too numerous to enumerate.

The IPCC's first and second assessments have proved to be a trove of knowledge and analysis. Nevertheless, some questions have been raised: Do the assessments consistently address the right questions? Do they reflect the best and most up-to-date scientific evidence and analysis? And can the analytical and factual insights in the assessments be translated into terms useful for decisionmakers in a timely way?

To address such questions, Bill Nordhaus has assembled a first-rate panel of ex post peer reviewers who address a number of themes. These themes include both general analytical issues, like the applicability of cost-benefit analysis to climate change and the proper analytical treatment of uncertainty, and specific topics related to climate change policies, such as the economy-wide costs of greenhouse gas control and the international benefits derived from abatement and adaptation measures. The volume also includes thoughtful comments on the reviews from other experts, including many of the original authors in the 1995 assessment. While the contributors are predominantly economists and other decision scientists, as befits the focus of the volume, they reflect a wide range of backgrounds and persuasions.

The chapters in this volume provide useful technical summaries of the conclusions from the 1995 socioeconomic assessment, making that material more accessible to the field, as well as thoughtful critiques of the information and reasoning used to derive some of these conclusions. The critiques and rejoinders should help to put the strengths and weaknesses of the IPCC conclusions in a clearer perspective, and to identify priorities for further research. In addition, the book is a useful tool for bridging gaps between analytical specialists on the one hand, and decisionmakers and their technical advisors in need of well-informed but accessible counsel on climate change issues.

Ten years after the establishment of the IPCC, and at the time that the IPCC is launching its third multi-year assessment of climate change issues, the material in this book may also help to draw attention to questions related to the IPCC process that require further debate. For example, how does one trade off the advantages and disadvantages of drawing upon recently written reports that are up-to-date but not yet peer-reviewed? Are there ways to make the analytical conclusions of the assessments more timely and accessible for decisionmakers, and what are the pros and cons of the practice of generating policymaker summaries that are subject to consensus approval by IPCC member states, not just the experts themselves? The need identified by Professor Nordhaus for increased understanding of climate change is likely to increase over the next decade. This book provides a solid platform for future progress to this end.

MICHAEL A. TOMAN
Director, Energy and Natural Resources Division
Resources for the Future

Preface

For most of its short and turbulent life as a political issue, discussion of global warming has been long on warm hearts and short on cool heads. This volume, which represents the results of a workshop held in Snowmass, Colorado, in August 1996, is the happy exception to that trend. In these pages, nineteen eminent analysts tackle the social and economic aspects of climate change and offer their views on the most difficult issues—cost-benefit analysis, the discount rate, the impacts and costs of climate change—as well as the framework issues of how to think about such long-run and uncertain problems.

The editor and contributors are grateful to the many people who helped turn the mountain of paper and disks into a polished manuscript. John Weyant and Susan Sweeney helped to host the original workshop and ensure that analysis was not interrupted by hunger or homelessness. The manuscript was tended from cradle to majority in New Haven by Marnie Wiss and Joe Boyer. Eric Wurzbacher, staff editor at Resources for the Future, and copyeditor Karen Coda buffed the rough-cut gems. Material resources were provided by the National Science Foundation through the National Bureau of Economic Research/Yale Program on International Environmental Economics. For want of these nails, this wisdom would be lost.

WILLIAM D. NORDHAUS
A. Whitney Griswold Professor of Economics
Yale University

Economics and Policy Issues
in Climate Change

1

Assessing the Economics of Climate Change

An Introduction

William D. Nordhaus

Understanding the science, economics, and policy aspects of global warming has proven one of the most exciting and challenging tasks facing the natural and social sciences over the last decade. Until the late 1980s, few people outside the geosciences had spent more than a fleeting moment reflecting upon the potential for or implications of global warming. Oil, natural gas, coal—these were the fuel for economic growth, the source of natural riches, and the cause of energy shocks. Global environmental change was still beyond the horizon.

The last decade has witnessed a sea change in attitudes and knowledge about many vital global processes, from the prospect of damage from ozone-depleting chemicals to the threat of deforestation and species depletion. But nowhere has research been so intense as in the field of climate change. The intellectual and policy challenges here have mobilized a small army of researchers to investigate every conceivable aspect of the problem. Although economics was a late entrant into the research process, today there are dozens of individual researchers or teams in all major industrialized countries looking at the different questions raised by the threat of climate change.

One of the special features about the phenomenon of climate change is its international dimension. In the language of economics, greenhouse warming is a global public good, or externality, over space and time in the sense that emissions in any one place affect the climate everywhere and for centuries to come. No single nation is responsible for the increasing amount of greenhouse gases, and no nation acting alone is likely to solve

1

the problem. Recognizing the global nature of climate change, nations organized an unusual process for reviewing the scientific and economic findings in this area. This process led to the report that is analyzed in the papers in this volume.

THE IPCC PROCESS

The Intergovernmental Panel on Climate Change, or IPCC, was established in 1988 by the World Meteorological Organization (WMO) and the United Nations Environment Programme (UNEP). All countries that are members of either the United Nations or the WMO are automatically members of the IPCC. Other international agencies and nongovernmental organizations often participate in activities of the IPCC, but they do not participate in adoption of reports. All reports are adopted by unanimous consent of the member governments.

As currently constituted, the IPCC has three Working Groups—on climate science; on impacts, adaptations, and mitigation; and on economic and social aspects of climate change. Every country that is a member of the IPCC is also a member of each of the Working Groups. The activities of the IPCC are directed by a Bureau of twenty-seven members. The Working Group Bureau includes two co-chairs: one from a developed country and one from a developing country or a country with an economy in transition. The subject of this volume is Working Group III, which examines economic and social aspects of climate change. This Bureau of Working Group III included two economists, a geographer, and a climate scientist.

The main activity of the IPCC is to assess the scientific literature related to climate change. The first assessment—primarily directed to the scientific issues—was completed in 1990 (IPCC 1990). The second assessment was launched at the IPCC meeting in November 1992, at which time the terms of reference of Working Group III that underlie the report analyzed here were determined. In May 1993 Working Group III held a workshop to develop a report outline and work program, and this work plan was accepted by the IPCC plenary in July 1993.

IPCC reports are written by scientific experts, usually four to twelve experts per chapter. The only restriction is that each writing team must include at least one expert from a developing country or a country with an economy in transition. The IPCC Secretariat solicits nominations for authors from the member countries as well as from international organizations and nongovernmental organizations. The Working Group Bureau selects the writing teams from the list of nominees based on their expertise, while also attempting to obtain reasonable geographic coverage.

Once the writing teams have been appointed, the members are free to invite other recognized experts to participate in the work.

The drafts of the IPCC reports and policymakers' summaries prepared by the writing teams are submitted to a technical review, administered by the Working Group Bureaus. The Bureaus identify technical experts and monitor the writing teams to ensure that they address the comments received by the technical review. Following the technical review, the drafts are submitted to government review, also administered by the Working Group Bureaus. The government review provides a formal opportunity for member countries to comment. Writing teams are expected to respond to comments on the basis of scientific merit.

Under the IPCC process for adopting a report, the Working Group must first approve, word-by-word, the Summary for Policymakers. IPCC guidelines specify that the summary should be consistent with the underlying technical chapters. The IPCC Working Group also decides whether to "accept" the technical chapters, but the content of the chapters remains the responsibility of the authors. Not surprisingly, negotiating unanimous word-by-word approval of Working Group III's Summary for Policymakers by more than 100 countries operating in six languages proved to be a difficult and time-consuming process.

Once approved, the report is submitted to the IPCC plenary for acceptance. Neither the Summary for Policymakers nor the underlying technical chapters may be changed by the IPCC plenary. All three volumes of the second assessment report were accepted by the IPCC plenary in Rome in December 1994. The three volumes were published by Cambridge University Press between June and August 1995 (IPCC 1996a, 1996b, 1996c).

THE SNOWMASS MEETING AND ASSESSMENT

The three IPCC reports of 1995 were widely recognized to be uniquely valuable reviews of current knowledge about climate change. These enormous volumes were produced by large teams of experts in the different fields and reviewed vast quantities of studies. At the same time, some of the results and interpretations were questioned. Debate about the validity of the conclusions, particularly about the scientific consensus, sometimes erupted into full-scale controversy. And even if the reports' conclusions did generally represent the current scientific consensus, that was no guarantee of scientific or predictive accuracy. For this reason, the contributors to the present volume thought it useful to stand back and take a careful look at the findings of the economic report to assess its reasoning.

Thus, a workshop was held in Snowmass, Colorado, in August 1996 with the purpose of analyzing in depth the economic and policy issues involved in the governmental consensus laid out in the Working Group III report. This workshop was organized under the aegis of the NBER-Yale Center on International Environmental Issues, funded by the National Science Foundation. For this meeting, recognized scholars who were not centrally involved in the IPCC chapters were asked to analyze the economic and policy issues involved in the assessment report. Each principal author prepared a review of a particular aspect of the report, and these reviews follow in this volume. We asked prominent scholars as well as lead authors of the relevant chapters of the IPCC report to respond to the analyses presented at the meeting. These subsequent comments also appear in this volume.

The chapters herein fall into two general categories. The first set of four chapters are general papers that review major analytical areas that are crucial for assessing the economic and social aspects of climate change. These include Policy Analysis for Decisionmaking About Climate Change by Professor Granger Morgan of Carnegie Mellon University; Equity and Discounting in Climate-Change Decisions by Professors Robert Lind and Richard Schuler of Cornell University; an overall appraisal of the issues involved in weighing costs and benefits in Applicability of Cost-Benefit Analysis to Climate Change by Dr. Paul Portney of Resources for the Future; and a discussion of alternative institutional design issues in Greenhouse Policy Architectures and Institutions by Professor Richard Schmalensee of MIT. These papers examine many of the central issues involved in framing the economic analysis of climate change.

A second set of papers examines specific topics necessary for the understanding of policies to cope with climate change. Perhaps the most intensively studied area involves the costs of slowing climate change. For this area, there are two reviews, one by Dr. Tom Kram of ECN Policy Studies, Netherlands, and the second by Professor John Weyant of Stanford University. Another critical building block is the issue of the benefits half of the cost-benefit analysis, which is discussed in Climate-Change Damages by Professor Robert Mendelsohn of Yale University. The final chapter, which reviews comprehensive models of the economics of climate change, is Integrated Assessment Modeling of Climate Change by Professor Charles Kolstad of the University of California at Santa Barbara.

The purpose of this introduction is to describe briefly the issues and conclusions in this volume and to provide some general background on the global-warming debate. The first section sketches the scientific background of the greenhouse-warming debate. The second section reviews the major analytical issues discussed in Part 1 of this volume. And the third section discusses the specific substantive policy issues raised in Part 2.

THE SCIENTIFIC BACKGROUND

What is the greenhouse effect? It is the process by which radiatively active gases like CO_2 selectively absorb radiation at different points of the spectrum and thereby warm the surface of the earth. Concern about the greenhouse effect arose because human activities are currently raising atmospheric concentrations of greenhouse gases (GHGs). The most important GHG is CO_2, most of which is emitted as a byproduct of the combustion of fossil fuels. There is no doubt about the accumulation of CO_2 in the atmosphere and little scientific doubt about the prospect of global warming over the next century and beyond if current trends in atmospheric concentrations of GHGs continue.

While the basic physics is well understood, the processes that govern the atmosphere and future climate change are extremely complex. There is therefore great uncertainty about the exact amount of warming, the timing of the climate change, and the regional distribution of the climate change. The uncertainties are enhanced because other atmospheric gases, particularly sulfates, interact with the GHGs. When incorporating uncertainties about both emissions and climate models, the latest round of IPCC projections indicates globally averaged, mean temperature increases of 0.5°C to 4.5°C from 1990 to 2100.

For nonscientists, one crucial question is how seriously to take this issue. As with most important matters, there are heated debates about the extent of greenhouse warming. It must be tempting for a politician with a time horizon of at most a few years, or a business with a time horizon of at most a few decades, to ignore a problem so full of controversy, whose impact will not truly be felt for a century or more. It must be tempting to hope it will go away or that someone will find a fundamental flaw in the science. This approach is ill advised. The basic science behind greenhouse warming is well established, and the basic projections represent mainstream science. There are vast uncertainties in the field, but these should not lead people to conclude that the issue can be ignored. A prudent strategy would be, at the least, to prepare for increasing concern and strong public policy measures in this area.

THE ANALYTICAL ISSUES INVOLVED IN CLIMATE-CHANGE POLICY

In light of the significant prospects for global warming, the next step is to consider the potential impacts of warming on human and nonhuman societies, the means and costs of slowing warming, and the most efficient and equitable policy approaches to tackling the problem. It is here that the Working Group III report enters the stage. Part 1 of this volume considers general analytical issues that frame thinking about or designing global-warming policies.

The Decisionmaking Framework

In many complicated areas the most difficult part of an analysis involves determining the appropriate ways to think about the diverse implications of different courses of action on human and nonhuman societies. As Granger Morgan shows in Chapter 2 on the decisionmaking framework, the challenges inherent in thinking about climate change are especially daunting. Our garden-variety economic analyses of issues such as zoning, drinking-water regulation, or highway expansion generally involve well-defined and relatively homogeneous groups and have relatively limited impacts extending over short time periods. For these issues there is a well-developed (although not always accepted) set of analytical tools and regulatory and legislative mechanisms—including such well-honed tools as cost-benefit analysis and decision theory.

It is natural to look to these tools when trying to sort out different approaches to the climate-change dilemma. Indeed, in later papers of this volume, they are extensively applied and analyzed, being, in a sense, the only analytical games in town. But their familiarity should not lead us into too-easy acceptance of their appropriateness, Morgan insists. The major difficulties with applying the standard analytical tools in this case are the complexity of the climate-change problem, its long time scale, and the fact that addressing it will involve the participation and require the assent of the various groups, generations, and "tribes" in different countries. Even if decisionmakers are persuaded that there is a sound case for taking costly steps to slow climate change, they must persuade legislatures, interest groups, and ultimately public opinion. Indeed, it is just this gap between the certainty of environmental groups and the indifference of the public that led to the schism between the administration and the Congress of the United States as that country formulated its climate-change policy in 1997.

One of the leitmotifs that runs through Morgan's analysis is the array of profound difficulties that arise because of the great uncertainties in climate change. Uncertainties not only raise genuine challenges for analysts, even more so, ambiguities pose daunting problems in educating a public that justifiably has trouble separating advertisement from information and science from advocacy. Uncertainties allow a leading automobile company executive to make the astonishingly misleading statement that most of the CO_2 emissions come from trees. Attempts to gauge uncertainties require complex expert-solicitation protocols, such as the one described by Morgan, but at best these can only generate subjective probabilities. And the traditional societal mechanisms for coping with risk are not well suited to handling the massive uncertainties and potential income redistributions that may arise from future global warming.

Discounting and Intergenerational Equity

One of the major issues that arises in analyzing the economic aspects of climate change is the discounting question. This issue, along with more general questions of equity, is discussed in the contribution of Robert Lind and Richard Schuler. It is easy to lose sight of the fact that alternative discount rates are intended to affect actual policy decisions. In the global-warming context, discount rates affect mitigation or adaptation policies regarding global warming, but they also have implications for the general economic environment.

A few words of background will illuminate the discussion. Beginning with the fundamentals, a discount rate is a pure number per unit of time that allows us to convert values in the future into values today. The most common form of discount rate is the nominal or money interest rate, which is applied to dollar values. When the nominal interest rate is corrected for inflation, we obtain the real interest rate, which converts future constant-dollar values into today's constant-dollar values (or future goods into present goods). Many writers confuse the discount rate on goods and services with the discount rate that applies to utility or general levels of well-being.

To understand the economics of real interest rates, economists often use the optimal-growth framework known as the Ramsey model. The Ramsey model derives the real interest rate from a combination of time discounting, the elasticity of marginal utility, and growth in consumption. Using these, it derives the famous equilibrium equation, $r^* = \alpha g^* + \rho$, where r^* = the equilibrium real interest rate, α is the elasticity of the marginal utility of consumption, g^* is the growth rate of per capita consumption, and ρ is the utility discount rate or the pure rate of time preference.

This equation—which is the basis for the discussion of discounting in the IPCC report and the Lind-Schuler critique—can usefully serve to explain the major schools in the debate. There are three major alternative approaches. One approach—sometimes called the descriptive approach—derives the discount rate for climate change from the left-hand side of the fundamental equation and proposes to use the same discount rate for environmental investments as for conventional investments. There may well be controversies about what numbers to use here (including the treatment of risk, taxes, and so forth), but this approach is clear in principle and tends to lead to relatively high discount rates.

A second approach—which has sometimes been called the prescriptive approach—rests on the argument that the discount rates that are actually found in markets are too high to be ethically justifiable because they discriminate against the future. This approach proposes using a lower and ethically justifiable discount rate. Proponents of this approach advocate discount rates as low as 1 % per annum. Because the discount rate is applicable everywhere in the economy, under this approach the economy should increase its overall savings rate sufficiently to drive the return to capital and the discount rate down to ethically justifiable levels.

Under a third scenario—the differential discounting approach—the descriptive discount rate is thought to be too high, but society is unable to generate the political will to reduce consumption. Policymakers would therefore apply special low discount rates to projects in ethically preferred habitats, such as the environment. In this case, conventional investments would continue to have the normal high discount rates, but the favored sectors would use the low discount rate for investments.

Lind and Schuler present a wide-ranging discussion not only of the issues discussed in the Working Group III report but also of the concepts underlying the economic philosophy of redistribution and intertemporal choice. One of their fundamental conclusions is that the entire discussion about discounting is largely misplaced. In their words, "any decision with regard to the mitigation of global warming is more fundamentally a decision about equity and the redistribution of welfare from the present and near-term generations to generations in the distant future and from the developed countries to less developed ones than it is a typical investment decision." From this fundamental premise they conclude that "there is no discount rate that can be based on either generally accepted principles of economic efficiency or generally accepted principles of equity."

The basic difficulty in determining the appropriate discount rate, in Lind and Schuler's view, is that the trade-off is not between income today and income in the future of a particular investor. Rather, the tradeoff is between *our* consumption today and *someone else's* consumption far in the future. In this view, we are not maximizing our net worth (as is the case in a typical investment decision) but deciding how much we in this generation would be willing to transfer to future generations. Indeed, because climate change is a pure public good, most of the benefits will go to other countries and are, in effect, foreign aid. So the first question in the global-warming debate can be viewed as how much intergenerational and international transferring we should engage in.

Having settled on the level of intergenerational and international transfers, the next question involves the form that these transfers should take. It is here that the opportunity cost of investments, or the return on investments of the descriptive approach, enters. For a given transfer rate, resources should go to those sectors where the (correctly priced) returns are highest. If the rate of return on convental investments are today high (say, 10 or 15% per year), then it would not be sensible to put resources into transfers or mitigation investments that yielded 1% or 2% annually. However, even this prescription is qualified, in Lind and Schuler's view, because of the impossibility of designing a set of transfers that would guarantee that an investment over a span of many generations would actually be carried out. Suppose that we decide that the most sensible investment is to start a large-scale forestation project to soak up carbon. We can plant the seeds, but we cannot guarantee that some future generation will not cut them down for firewood or fancy houses.

Despite these reservations, Lind and Schuler offer some helpful conclusions about the role of discounting in analyses of the climate-change issue. They do not recommend throwing all discount rates out the window. Rather, they point out that an "opportunity-cost test" is a useful check on proposals. That is, we should always ask whether our investments and transfers can be deployed in a way that makes all generations (or countries) better off. But they urge modelers to avoid oversimplifying their results by showing one single number (like the discounted value of utility or consumption) which hides too many details and assumptions. Rather than having flows discounted back to a single number, they would prefer to see the time paths of consumption, utility, and the environmental variables so that readers of these studies can form an independent judgment about the values of different policies.

Role of Cost-Benefit Analysis

Paul Portney of Resources for the Future examines the question of the applicability of cost-benefit analysis (CBA) in the context of climate change. Many analysts question the applicability of CBA in this context because climate change is the "mother of all problems." The IPCC review group points to many features of climate change that are, in their view, of unprecedented complexity. These involve the irreversibility and complexity of underlying geophysical processes, the global nature of the issue, and the paucity of the data on costs and benefits. Portney takes issue with this view, arguing that similar complexities are standard fare for those who do CBA on environmental problems. He points to issues like local air pollution or regulation of chlorofluorocarbons as examples of problems that contain similar analytical complexities.

Portney goes on to discuss problems that are sometimes underemphasized in analyses of CBA. Most of these involve valuation of costs and benefits. One pervasive difficulty is in estimating costs, which would entail evaluating general-equilibrium models to yield appropriate figures. Even more difficult in most cases is the evaluation of benefits, or damages avoided, by slowing emissions. Portney illustrates this problem with the example of estimates of willingness to pay (WTP) for reducing premature mortality. The preferred current approach among economists is to use a WTP based on the amount that individuals "charge" to accept higher risks in workplaces and in other circumstances. Estimates for the United States used in regulatory analyses tend to cluster around $4 million for a statistical life saved. (Even this number raises many statistical and substantive questions.) Portney asks whether we are comfortable using this methodology when doing a CBA that involves people with vastly different income levels and therefore is likely to involve very different values of a statistical life. How should we conduct our CBA if our methodology leads to the conclusion that the WTP for an Aspenite life is $40,000,000 while that of a Bengali is $40,000? Would our international negotiations consider for a moment policies that would load the premature mortality on low-income countries because this was "cost effective?" The reaction of most analysts who peek in this door is to close it very quickly and move on.

In his final section, Portney raises a different interpretation of the appropriate role of CBA by suggesting—as a thought experiment—a global referendum on climate change. He suggests that we think of CBA not as a mechanical adding up of costs and benefits but rather as a way of asking members of the current generation (either as individuals or governments) how much they would be willing to pay to reduce the threat of future climate change. In essence, we should be asking ourselves how much we are willing to provide in transfers to the domestic or foreign indigent or how much we would pay to set aside a tract of land for wilderness instead of enjoying the benefits of mining the oil or gold under that land. In the climate-change perspective, we would be asking how much current consumption we are willing to set aside to reduce the chance that mundane or catastrophic consequences will occur in the future. This thought experiment is conducted in quite a different spirit than the standard CBA, which asks how much we are willing to reduce our private consumption to increase public consumption of the environment or of road use by the current generation. Portney's conception is very similar in this respect to Lind and Schuler's view of the appropriate way to evaluate the discount rate and intergenerational equity.

Institutional Issues

In environmental affairs, problems attract solutions. The response of nations to the threat of global warming has been to join together to find ways of reducing emissions. The approach followed has been that of the

Framework Convention on Climate Change (FCCC), which issued from the Rio Summit of 1992. (This approach was further defined by the set of decisions made in Kyoto in December 1997 after the analyses by the IPCC reports had appeared and after the papers in this volume were written.) Under the FCCC, high-income nations (plus the former Soviet Union and Eastern European countries) committed to limit their concentrations of GHGs to 1990 levels. This commitment left open almost all the important questions, such as the environmental, economic, and political aspects of such a commitment. These issues and the review of these issues by Working Group III of the IPCC are analyzed by Richard Schmalensee in his paper, Greenhouse Policy Architectures and Institutions.

The basic point of Schmalensee's analysis is that the IPCC review underestimates the difficulties involved in building the institutions required to have an effective and efficient instrument to slow greenhouse warming. The first omission of the review, according to Schmalensee, is that it pays insufficient attention to the development of long-term policies and institutions. Most discussions seem to envision policy as if it were a once-and-for-all decision about the level and distribution of emissions reductions. Rather, policymaking in this context should aim to develop an institution that has sufficient flexibility to make multiple midcourse corrections as science or economic analysis uncovers new evidence on the evolution of global warming. At an elementary level, the Rio and Kyoto strategies are poorly designed because policies should focus on consequences (like climate change) rather than on intermediate indicators (like emissions). The current approach is also too inflexible. For example, recent climate-modeling runs have uncovered a potential threshold in the reaction of North Atlantic deep-water circulation to the level and pace of climate change. Any sensible institution should be able to incorporate such evidence (after appropriate analysis and review) into climate-change policy. The Rio and Kyoto approaches, which focus on reducing emissions relative to a historical emissions base, are very poorly suited to making such adjustments.

A second key analytical deficiency in both the report and much current analysis, in Schmalensee's view, is the assumption that a "climate czar" will make the "correct" decisions about climate-change policy. Schmalensee points to the necessity of devising strategies that will be in the interest of the several nations who not only sign on but also take costly steps to reduce emissions. The political difficulties are particularly profound if, as Schmalensee and the report both believe, "reducing CO_2 emissions substantially relative to trend…would likely involve annual costs on the order of several percent of world income." He notes that the difficulties are compounded by the absence of an established procedure for measurement and monitoring. The need to induce participation, monitor and ensure compliance, and prevent free riding poses awesome difficulties.

The current approach to climate-change policy has sometimes been called "deep, then broad" to indicate that it would begin with deep cuts among a narrow group of rich countries and then follow with broad participation at some time in the future. In his final section, Schmalensee suggests that alternative architectural designs—particularly, a "broad, then deep if necessary" strategy—would ultimately prove more effective. The alternative "would place less stress on near-term emissions reductions, which are of relatively little importance over the long haul, and would concentrate instead on developing institutions to ensure broad international participation in emission abatement, which is essential to any effort." Such an approach would focus on a hybrid of tax and quantity measures as well as on ex ante policy measures (such as taxes) rather than ex post results (emissions). The hybrid approach would emphasize information, education, research, measurement, and consensus building in the short run rather than trying to cram large and costly emissions reductions down the throat of poorly informed and largely unwilling citizens in high-income countries. This alternative design is one that clearly warrants careful thought and study by analysts as well as policymakers.

SPECIFIC TOPICS IN CLIMATE-CHANGE POLICY

In addition to the more generic issues that would arise for almost any important long-term and global public good, climate change raises a number of specific questions. Three central economic questions reviewed in this volume involve the costs of slowing climate change, the damages from not slowing climate change, and a new analytical approach—called integrated assessment—that brings the different pieces of the problem together.

Abatement Costs

The first set of issues involves the costs of slowing climate change. According to current understanding, climate change can be slowed by reducing the growth of atmospheric GHG concentrations through conventional means, such as reducing CO_2 emissions, or by taking more unconventional steps, such as planting trees or undertaking geoengineering experiments. The IPCC and this volume focus largely on the conventional approach (although a 1992 study by the U.S. National Academy of Sciences included a more complete list of options [National Academy of Sciences 1992]).

The papers by Tom Kram and John Weyant review the troops in this area. Two major points emerge from Weyant's survey. First, the IPCC and the studies that Weyant cites agree that the costs of current climate-change proposals are extremely large in the context of current economic and environmental policies. For example, to stabilize emissions at 1990 levels over the next fifty years would require a few percent of world output. If world output grows at 3% over this period, this would be a cumulative expenditure level on the order of $100 trillion. The program would be larger than all the rest of today's world environmental programs put together. In short, there are really big stakes in this game.

The second major point in Weyant's paper is that these estimates are subject to large uncertainties—indeed, his subjective standard error of the cost of abatement is approximately as large as his central estimate. The uncertainties arise partly because of intrinsic difficulties, such as uncertainty about the evolution of population and technology, and partly because of unresolved modeling issues, such as the presence of organizational or individual slack and the consequent potential for large, low-cost, energy-saving improvements. Particularly interesting is Weyant's comparison of three alternative approaches to modeling—the economic, engineering, and social-psychological. He shows that these different conceptions give radically different interpretations of history and prognoses for the prospects of energy saving in the future.

Weyant concludes with a long list of areas for further research. I will mention one here—the double-dividend controversy—to indicate the treacherous nature of the cost calculations. Although most proposed mechanisms to reduce emissions involve some kind of regulatory mechanism, they have quite different impacts upon government revenues. Those plans that are favored by economists include carbon taxes and auctioned emissions quotas—both of which raise revenues, in some cases substantial revenues. For example, the 1997 U.S. government analysis of its own proposal indicated that it could easily raise $100 to $300 billion in revenues if the emissions limitations were sold rather than allocated. The fiscal impact comes in two steps. First, conventional economic analysis indicates that, because of the higher prices and consequent lower real wages and lower real incomes that would result from emissions limitations, the deadweight loss involved in paying for existing government programs would increase. Second, if these revenues were returned to consumers (or recycled) by lowering other distortionary taxes, some and perhaps all of these distortionary impacts might be eliminated.

Here is where the "double-dividend" hypothesis that Weyant mentions enters. Much enthusiasm was generated by the prospect that "green taxes," such as carbon taxes, might reduce the inefficiency of the current tax system. The idea was essentially that a carbon-tax-cum-rebate plan would achieve not only the environmental benefit (reducing emissions and the consequent damage from climate change), but also the second of the fiscal impacts just mentioned: reducing the inefficiency of existing taxes. After much careful work by tax economists, it was shown that this hypothesis was essentially false. Its proponents overlooked the first of the two fiscal impacts—that the price-raising effects of the emissions limitations would themselves increase the deadweight loss of existing taxes. Fiscal specialists now generally agree that—absent particular features of energy taxes—the fiscal effect of substituting carbon taxes for other distortionary taxes is roughly neutral.

This finding does have one important implication, however. Any plan that includes emissions limitations without raising revenues (that is, one that restricts emissions and raises energy prices with the revenues flowing to, say, grandfathered interest groups) would have distortionary effects. Indeed, some calculations indicate that an optimized emissions strategy, which assumes fiscal neutrality but in which the government "loses" the revenues, might actually have negative net impacts when the fiscal impacts are included. This example shows that considerations generally thought to be secondary can easily become of primary importance if the architecture of the plan is not carefully designed.

The paper on mitigation costs by Tom Kram examines both the analytical framework in Chapter 8 and the modeling results in Chapter 9 of the Working Group III report. Kram notes that there is tremendous heterogeneity among the different models and approaches. For example, the IPCC surveyed more than two dozen comprehensive studies to provide estimates of the costs of abating GHGs. Kram points out that the models are not equally well developed—the close scrutiny required to determine which are the most appropriate models was unfortunately missing from the IPCC's review. He is particularly concerned about the oversimplified technological assumptions and poor documentation in many models and notes that many models were neither designed to be used for nor readily adapted to the questions at hand.

In exploring different approaches, Working Group III makes much of the difference between "top-down" and "bottom-up" models (often characterized as economic pessimism and engineering optimism). Kram notes that while this distinction had much validity in the early days, considerable hybridization has taken place as models adopt the best features of both approaches. Top-down economic models increasingly incorporate engineering details while bottom-up engineering models have begun to recognize that the demand for energy services responds to prices.

In examining work in this area, Kram lists three important questions that studies must answer if they are to be useful to policymakers:

- How costly is reduction of greenhouse-gas emissions?
- How uncertain are the cost estimates?
- What differences across countries can be identified?

In Kram's view, Working Group III only touched on these issues. One of the major shortcomings of current research, Kram emphasizes, is the sparsity of comprehensive independent estimates of the cost of abatement for developing countries and economies in transition. Estimates for the latter are particularly problematic given the rapid rate of structural change.

In the final section, Kram discusses a number of issues overlooked in the IPCC survey. Two points are particularly valuable for assessing mitigation-cost studies. The first involves the measuring of "no regrets" or negative-cost options. Using engineering-based bottom-up models, many analysts have found that considerable reductions of CO_2 emissions can be achieved with negative or very low costs. In fact, the modeler here is assuming a cost-free and largely unexplained improvement in the efficiency of the energy system. The example of refrigerators shown in Figure 3 of Kram's paper illustrates how the cost function depends critically on the assumed path of best-practice technology.

The other interesting point is illustrated in Kram's Figure 5, which shows the least-cost mitigation profile for different technological options (conservation, nuclear, renewals, and CO_2 removal). Such comparisons would be impossible without detailed process models, which allow a rich technological detail. One striking result of Kram's analysis is the attractiveness of CO_2 removal, which has been little discussed in most countries.

Impacts

What are the likely impacts of projected climate changes over the next century and beyond? What are the impacts on different regions and sectors, on the market economy, on nonmarket activity, and on ecosystems? Are there significant differences in impacts for high-income and low-income countries? How likely is it that climate change will have catastrophic consequences? We need answers to these central questions if we are to design sensible public policies. Estimating damages is not only crucially important but, of all the areas in the climate-change debate, this has proven the most difficult and controversial.

Robert Mendelsohn reviews the IPCC approach as well as an alternative assessment. WGIII Chapter 6 concentrates on providing not only a review of the impacts but also an estimate of the monetary value of the impacts. Because this is such a controversial area, it is worth reiterating the reason why economists monetize the benefits. The standard economic approach to environmental problems today is to determine policies that balance costs and benefits. This requires not only estimating the monetary costs of abatement policies (which was undertaken in the WGIII chapters reviewed by Weyant and Kram) but also estimating the monetary benefits.

Mendelsohn commends the authors of Chapter 6, on social cost, for performing the important task of estimating the total economic impacts of climate change. He also defends the controversial decision of the authors to weigh in on the debate over valuation of human life as a crucial part of any valuation exercise. Chapter 6 presents a number of economic estimates and carefully reviews many of the analytical and empirical difficulties posed by damage estimation. The chapter concludes that the climate change resulting from a doubling of CO_2 would lead to global damages of 1.5% to 2.0% of world output. These damages would fall more heavily on developing countries (with a range of estimates of 2% to 9% of outputs) than on developed countries (with a range of estimates of 1.0% to 1.5% of outputs). The chapter also estimates that the marginal damage of a ton of current carbon emissions. The estimates from integrated assessment models (discussed in the Kolstad paper in this volume) ranges from \$5 to \$12 per ton of carbon when using a 5% discount rate on goods and services. Other estimates, based on lower discount rates and alternative approaches, range from \$5 to \$125 per ton of carbon.

Mendelsohn has a number of criticisms of the social-cost chapter. In brief, he concludes that the chapter has an inadequate appreciation of the stock nature of the climate-change problem of CO_2 emissions, that it systematically underestimates the importance of human societies adapting to climate change, that it includes unfavorable impacts while ignoring beneficial impacts, and that it gives insufficient attention to the need to base valuations on sound empirical evidence. Mendelsohn cites several examples of these shortcomings; for example, the report overlooks the significant potential benefit to agriculture and forests of CO_2 fertilization, it assumes that storm damage will increase even though the scientific assessment of the IPCC was agnostic on the impact of climate change on storminess, and it uses an inappropriate methodology that ignores the role of adaptation in its assessment of the impact of climate change on human health.

In summary, Mendelsohn argues that "the estimates [of the damage from climate change] in the chapter should be treated as upper bounds of what might happen. What will actually happen is highly uncertain, but the expected value of impacts is not likely to be as harmful as the chapter implies." In other studies, Mendelsohn and his colleagues have made an alternative assessment of the likely impact of climate change. He summarizes this assessment as finding that "the world economy may even gain a small benefit from mild warming."

Whatever their assessments of the most likely path of impacts from climate change, all those who have studied this area agree that our estimates are at this time highly uncertain, and we have at best an imprecise picture of the human societies on which the changing climate will act. Moreover, outside of a few sectors (primarily agriculture and sea level in the United States), we have at best rudimentary studies of the impacts of these highly uncertain scenarios. Indeed, the largest uncertainties pertain to exactly those regions where the impacts are expected to be the most significant—low-income, tropical regions. All would agree that much research remains to be done to improve the confidence in our estimates of the impacts of climate change.

Integrated Assessment

The final paper, by Charles Kolstad, reviews the field of *integrated assessment modeling*, or IAM. Although definitions of IAM differ, the general nature of such models is that they are numerical computational models of complex processes that cross disciplinary lines and address important public policy issues. For the most part, IAMs are integrators of basic disciplinary research; they do not do the underlying research on impacts or economics or climatology but rather attempt to incorporate the disciplinary insights into their models. There have been a number of examples of IAMs over the last thirty years, including those addressing energy and environmental policy (such as aspects of energy independence or acid rain). Over the last decade, there has been growing attention to building and using IAMs for analysis of the implications and policy aspects of climate change.

Kolstad and the IPCC review chapter divide the current crop of IAMs in climate change into two species: policy evaluation and policy optimization. Policy-evaluation IAMs trace out the implications of different scenarios or policies for a wide range of variables such as emissions, concentrations, global and regional climate variables, ecological impacts, and socioeconomic impacts. Different models emphasize different aspects of this complex chain, but—because they are generally recursive simulation models—policy-evaluation models can handle extremely large systems. The large policy-evaluation models are similar to giant general-circulation climate models which can easily take a few months of supercomputer time to make a single run.

Policy-optimization models, on the other hand, usually focus on determining an "optimal" or at least a "good" policy response. Most policy-optimization models are structurally simple but computationally complex, dynamic cost-benefit models that integrate all the different elements of the logical structure into a single model. Such optimization models tend to be relatively small because of the mathematical limitations on maximizing extremely complex numerical problems. Kolstad concludes with a controversial finding: "In reviewing the fundamental knowledge of climate-change policy that has emerged from IAMs, it is striking that nearly all of the results have come from the so-called policy-optimization models, the top-down economy-climate models."

In his review of the different IAM families and the species within each family, Kolstad identifies a number of robust results. Three stand out:

- Current assessments determine that the "optimal" policy calls for a relatively modest level of control of CO_2
- The timing of emissions is economically important, and in particular it seems efficient to "backload" major emissions reductions if the objective is to limit concentrations of greenhouse gases or temperature change
- The discount rate matters a lot for the optimal control rate—discount rates found in capital markets today would lead to the modest control rates of the first bulleted item while low discount rates would lead to significantly higher control rates.

There was surprisingly little dissent from these rather strong conclusions among participants and discussants of the chapter.

Kolstad also reflects on some of the shortcomings of and challenges for IAMs—issues that are on the agenda of current and future research, as seen in the IPCC report and in the more general policy discussion, and that represent a good summary of the major unresolved issues in climate-change research. Some of the problems of IAMs are general problems originating in disciplinary uncertainties. The most important example is that our inadequate understanding of the impacts of climate change reflects uncertainties about the quantitative impacts as well as the valuation of those impacts. This is not, in Kolstad's view, a problem of IAMs but a problem of the underlying physical and social sciences. Similar challenges arise in improving the cost and technological estimates, the representation of learning and technological change, and the resolution of climate-change predictions such as the regional impacts and the rate of change of climate.

One area on which Kolstad focuses in his review is the treatment of uncertainty and learning. Many analysts today believe that the climate-change problem is one of coping with the surprises rather than with the most likely impacts of increasing concentrations of GHGs. Kolstad notes that while some work has been done in this area, "there is little consensus on the implications of large damage/low probability events for climate-change policy." He notes that progress will require two developments. First, much more work must be done to identify potential catastrophes and quantify their probability and consequences. Analysts then need to determine how to both incorporate the catastrophes into the IAMs and examine the consequences and potential policy responses.

In one sense, IAMs attempt to incorporate the accumulated wisdom about climate change of the natural and social sciences into a single integrating structure. To this observer, the most striking feature of this field is its dynamism. A decade ago, not a single IAM existed. Over the last decade—reflecting both intense public policy interest and tremendous attention in the different disciplines—literally dozens of teams have developed different models to investigate a wide variety of issues. Moreover, the results and the mind-set underlying the IAMs have begun to influence the policy discussions. Whereas in the early 1990s the only policy question under discussion in diplomatic circles was whether to set constant or slightly lower constant paths for GHG emissions, today a wide range of policies is under discussion. The importance of timing and location of emissions ("when and where flexibility") is becoming apparent because of the consensus of models about the efficiency gains from such flexibility. The concept of emissions-trading protocols, introduced by the United States in the summer of 1996 and embedded in the Kyoto Protocol of 1997, traces back to studies of the gains from trading in IAMs. Much remains on the agenda of modelers, but accomplishments to date in this complex area are already noteworthy.

FINAL ASSESSMENT

In the end, the papers in this volume help put the Working Group III report in historical and political perspective. A reading of the report and the papers in this volume leads the present author to a few conclusions. First, the IPCC process has in one sense been remarkably successful. It has succeeded in producing a summary of the state of the art in a field that is highly fragmented. Although the field is moving rapidly, people interested in the area would be well rewarded by a careful study of the Working Group III report. It is an unparalleled compilation of studies, including full references and comparisons of different approaches. As a document, it is probably not terribly useful to governments because it is too long and technical to provide guidance on climate-change policy. But it will be of great help to technicians and researchers, particularly those who are just finding their way into this complicated field.

Second, a number of chapters in the report are superb essays on their subjects. While they are not without flaws, we can point to the chapters on discounting, integrated assessment, and cost comparisons as ones that not only review the literature but also organize the issues in a new and thoughtful manner. These chapters provide useful summaries of the current state of thinking on the subject and contain summary information on current findings of empirical research.

Third, some of the chapters have proven to be less successful, particularly the ones on social costs and on policy architecture. To some extent, the difficulties in these chapters reflect weaknesses in the underlying literature. For the issue of social costs, this weakness mainly reflects the difficulties inherent in research on the impacts of climate change and the paucity of sectors (mostly ones in the high-income countries) where solid empirical studies exist. In addition, in the area of damages and impacts perhaps more than in others, there is a strong tendency to see the cloud behind every silver lining; much analysis, particularly by environmentally oriented researchers, has focused on the damages and ignored the potential benefits of climate change. The weakness in the chapter on policy architecture probably reflects the fact that virtually no serious thinking has taken place on the issue of how to design institutions to efficiently manage global economic public goods. In short, the weaknesses in the report largely mirror immaturity in the underlying scholarly literature, which also suggests that much more research needs to be focused on these two areas in the future.

The Working Group III report will probably not be bedside reading for any but insomniacs. But for those seriously interested in the economic dimensions of climate change—always recalling that the field is moving rapidly and studies can easily be superceded by new results—the IPCC report stands as a premier reference book on the subject. The fine essays in the present volume can serve as useful companion pieces to help readers and policymakers understand the strengths and weaknesses of the different approaches and point researchers toward areas that need further analysis.

ACKNOWLEDGMENTS

The section in this paper entitled The IPCC Process draws heavily on an unpublished description generously provided by Dr. Erik Haites, one of the chief organizers of Working Group III of the IPCC.

REFERENCES

IPCC (Intergovernmental Panel on Climate Change). 1990. *Climate Change: The IPCC Scientific Assessment.* Edited by J. T. Houghton, G. J. Jenkins, and J. J. Ephraums. New York: Cambridge University Press.

———. 1996a. *Climate Change 1995: The Science of Climate Change. The Contribution of Working Group I to the Second Assessment Report of the Intergovernmental Panel on Climate Change.* Edited by J. P. Houghton, L. G. Meira Filho, B. A. Callendar, A. Kattenberg, and K. Maskell. Cambridge: Cambridge University Press.

———. 1996b. *Climate Change 1995: Impacts, Adaptations, and Mitigation of Climate Change: Scientific-Technical Analysis. The Contribution of Working Group II to the Second Assessment Report of the Intergovernmental Panel on Climate Change.* Edited by R. T. Watson, M. C. Zinyowera, and R. H. Moss. Cambridge: Cambridge University Press.

———. 1996c. *Climate Change 1995: Economic and Social Dimensions of Climate Change. The Contribution of Working Group III to the Second Assessment Report of the Intergovernmental Panel on Climate Change.* Edited by J. P. Bruce, H. Lee, and E. F. Haites. Cambridge: Cambridge University Press.

National Academy of Sciences. 1992. *Policy Implications of Greenhouse Warming: Mitigation, Adaptation, and the Science Base.* Committee on Science, Engineering, and Public Policy. Washington, D.C.: National Academy Press.

PART 1
Major Analytical Issues

2

Policy Analysis for Decisionmaking about Climate Change

M. Granger Morgan

In addressing a problem as complex as climate change, policy analysis is essential. Without it, decisionmakers may not even know how best to frame the policy choices they must make, let alone develop informed preferences on which to base those choices. Policy analysis is a difficult business. There are many available tools, but knowing when each is appropriate, how they should be combined, what assumptions underlie each, and how much or how little formal analysis to perform given the uncertainties and other characteristics of the problem at hand requires technical expertise, experience, and well-developed professional judgment. Thus it is entirely appropriate that in planning the report of the IPCC Working Group III (WGIII), the organizers chose to include Chapter 2, "Decision-Making Frameworks for Addressing Climate Change." In this paper I review and discuss that chapter.

The first part of this paper reminds readers that the authors of WGIII Chapter 2 had a very limited climate-specific research base available for their review and warns against overgeneralizing about the utility of social science and policy analysis for the problem of climate change on the basis of the Working Group III review. Next I provide a brief summary of the contents of Chapter 2 and raise several concerns about the chapter's treatment of the issues before concluding with point-by-point commentaries on the chapter's conclusions about decision analysis and negotia-

M. GRANGER MORGAN is head of the Department of Engineering and Public Policy and Lord Chair Professor in Engineering at Carnegie Mellon University.

tion frameworks. I conclude the paper by arguing that the basic assumptions that underlie many of the quantitative analytical methods now being employed by researchers working on the socioeconomic dimensions of climate change need critical reassessment. A critical examination of these tools, and a discussion of the research needed to produce more appropriate tools, should be a central preoccupation of the next IPCC review of the economic and social dimensions of climate change.

IT IS HARD TO WRITE A REVIEW
WHEN THERE IS NOT MUCH LITERATURE

Working Group III was charged with summarizing the current state of the scientific literature on the economic and social dimensions of the climate problem. In evaluating their performance, and comparing it to the reports of Working Groups I and II, it is important to remember that the physical science of climate change has been a focus of serious research for something like half a century. During that time the field has received billions of dollars of research support, collected vast amounts of geophysical data, and constructed some of the world's most computationally demanding computer models. The study of ecological impacts from climate change has received far less attention. Nevertheless, serious work has been in progress for several decades, and support has certainly totaled some hundreds of millions of dollars.

In contrast, with a few exceptions, serious work on the socioeconomic dimensions of climate change, and on performing, developing, and evaluating policy analysis on the climate problem, has been in progress for only a decade. Total support has probably not exceeded a few tens of millions of dollars. Not surprisingly the product of Working Group III is a good deal leaner than the products of Working Groups I and II. There is, of course, a much larger general literature in social-science research and policy analysis on which the authors of Working Group III could draw. Later I will argue that the authors of Chapter 2 did an inadequate job of drawing upon that general literature.

Until quite recently, a physical-science perspective has dominated the design of climate-research programs, as it has the coordination of the IPCC review. A significant fraction of physical scientists believe that the fields of social and behavioral sciences and of policy analysis are very limited in their ability to make useful contributions to a problem like climate change, assuming that once the geophysics has been done correctly, other aspects of the problem will be rather straightforward. This belief can be illustrated with data we recently collected during a set of detailed interviews with sixteen leading U.S. climate scientists (Morgan and Keith

1995). In addition to posing a wide variety of questions about the nature, extent, and limitations of current scientific understanding of climate science, we obtained a series of judgments about what these experts believed were appropriate levels of research support. After constructing a detailed zero-based budget which allocated a billion dollars per year to research in climate-related geophysics for each of the next fifteen years, the experts were asked to indicate how much they would consider to be "socially appropriate" to spend in a number of other climate-related areas. Table 1 summarizes the results. While the responses of several of the climate scientists would advocate a significant increase in research support in the areas of "socioeconomic impacts and adaptation" and "integrated assessment," a number of them believe that it would be socially appropriate to spend only 1% or 2% as much in these areas as in geophysics.

When they read the IPCC report, many physical scientists will probably ignore the very different levels of previous effort that underlie the literatures available to the authors in the three Working Groups. Instead, they will take the obvious limitations of the report of Working Group III as a confirmation of their previous belief and conclude that policy analysis and research in the social and behavioral sciences is pretty limited in what it can contribute.

BASIC APPROACH OF CHAPTER 2

The authors of WGIII Chapter 2 discuss "possible decision-making frameworks related to climate change" (IPCC 1996, 57). They begin with "a review of some of the unique features of the climate change problem and their implications for decisionmaking" (57). Next they discuss uncertainties, particularly those associated with impacts. They identify as particularly important the long time horizons that the climate problem involves. Then they introduce and discuss two possible models for decisionmaking: decision analysis and negotiation. The chapter concludes with a discussion of implications for national decisionmaking under the Framework Convention on Climate Change (FCCC).

The two models considered in Chapter 2, decision analysis and negotiation, both explicitly involve identifiable decisionmakers who consider alternatives and make a choice. As Schelling (1978) has illustrated, in many social processes, important macro-outcomes do not result from such a conscious deliberative process. Rather, in many circumstances (including such simple situations as choosing which neighborhood to live in, where to sit in an auditorium, or whether to keep attending a seminar), the simple decision rules used by individual decisionmakers can col-

Table 1. Estimates by Experts of Socially Appropriate Research Investments in Different Parts of the Climate Problem.

Research area	1	2	3	4	5	6	7	8	9	10	11	12	13	14	15	16	av
												Expert number					
Human emissions	100	50	100	20	50	30	10	50	60	10	100	200	50	20	100	10	60
Geophysics	1,000	1,000	1,000	1,000	600	1,000	1,000	1,000	1,000	1,000	1,000	1,000	1,000	1,000	1,000	1,000	–
Ecosystem response	200	100	100	500	100	60	10	50	300	5	100	300	300	40	200	10	130
Socioeconomic impacts and adaptation	300	40	100	200	40	30	10	100	300	10	100	200	200	10	200	10	120
Strategies for abatement	2,000	100	400	500	100	50	800	100	5,000	–	100	500	20	20	300	30	630
Strategies for geoengineering	0	0	20	100	40	30	200	100	10	0	100	20	100	4	30	0.2	47
Integrated assessment	20	20	10	100	6	10	10	50	60	0.5	100	100	100	20	10	2	38

Note: Sixteen leading U.S. climate scientists were asked by Morgan and Keith (1995) to indicate the annual level of research support that they would consider "socially appropriate" in a number of different research areas assuming that climate science was being supported at a level of $1 billion per year. The table displays the results obtained, reported in units of millions of U.S. dollars per year.

lectively produce quite unexpected, and often undesired, collective outcomes.

In my view the literature on decisionmaking can be roughly divided into four categories, as shown in Table 2. In several cases the same literature appears in both descriptive and prescriptive cells because both versions exist in the literature. Chapter 2 has mainly dealt with the two italicized entries in this table. I believe that a number of other entries, especially some of the collective-descriptive entries, are actually more relevant to social decisionmaking on issues of climate change.

The authors of WGIII Chapter 2 do correctly point out that "there is no single decisionmaker in climate change" (IPCC 1996, 58). However, having made this observation, they then largely adopt the perspective of the FCCC, which views the world as made up of a group of one hundred-odd national parties, each of which behaves as a unitary rational actor that can exercise effective control over the activities that occur within its boundaries. Allison (1971) has demonstrated that this assumption can miss, or fail to adequately explain, a variety of important behaviors that derive from such social processes as bureaucratic politics and the varying cultures and standard operating procedures of different agencies. In the case of climate change, the situation is yet more complex because, while

Table 2. Possible Categorization of the Literatures on Decisionmaking.

	Individual	*Collective*
Descriptive	Microeconomics Behavioral decision theory (cognitive limitations: satisfying, prospect theory, etc.)	Markets in microeconomics Game theory (including prisoners' dilemmas) and negotiation theory Organizational behavior Bureaucratic politics Political theory International relations theory
Prescriptive	Microeconomics Other forms of utility maximization (benefit-cost, *decision analysis*, multi- attribute utility theory, etc.) Ethical, cultural, and other prescriptive rules	Game theory *Negotiation theory* Welfare economics (Pareto optimality, etc.) Legal and other procedural models Organizational behavior Bureaucratic politics Political theory International relations theory

Note: The authors of WG III Chapter 2 chose to address only selected aspects of the two literature categories that have been italicized: decision analysis undertaken from the perspective of a single rational actor and negotiation among a set of rational actors.

governments set general energy policy, most of the decisions that directly determine future emissions and loadings are *not* made by government decisionmakers. Rather, they are made by thousands of separate companies and millions of individual consumers.

In principle, national governments exercise significant decision control over the technology choices and emissions patterns of the firms and individuals that operate within their boundaries. In practice, particularly in most representative democracies, the freedom of action that governments have is severely constrained by political circumstances and by pressures from important interest groups. For example, while many knowledgeable observers of the U.S. energy scene believe that a gasoline tax of a dollar or more per gallon would send market signals that would have a number of beneficial effects on the environment, on new, cleaner transportation technology, and on national security, public commitment to low gasoline prices prevents any serious consideration of such an option.

In his book *Agendas, Alternatives and Public Policies*, Kingdon (1984) has advanced a behavioral model in which "streams" of problems, policies, and politics proceed largely separately from each other but occasionally become aligned—because of events in either the political or the problem stream—to form a "policy window." Kingdon argues that "open windows present opportunities for the complete linkage of problems, proposals, and policies, and hence an opportunity to move packages of the three joined elements up on decision agendas" (213). In the context of climate, we have recently seen how unusual circumstances, such as a series of hot summers, or intense storms, can influence decision agendas.

The Kingdon model suggests that national policies toward climate matters will evolve through a series of discrete, somewhat random, steps, which often will not be in synchrony from nation to nation. Thus, a description in terms of decisions by single national rational actors, or even in terms of negotiation among a number of such actors, misses much of the essential character of the sociopolitical process.

While the problem of climate change is accompanied by much greater uncertainty, it has some of the same attributes as the growing problem of entitlements in the budgets of the United States and other OECD countries. These governments have clearly demonstrated that, in the face of short-term political realities and pressures from interest groups, they have great difficulty taking action to address a clearly predictable, but gradually building, problem.

In writing about the FCCC, Chapter 2 argues that "the Convention is, first and foremost, a framework for collective decisionmaking by sovereign states" (IPCC 1996, 58). While officially this is true, the reality gets more complicated. Clearly the motivations of the key participants in this process are mixed. A few participants are working seriously to address and solve climate problems. Many more, both governments and interest groups, are pursuing shorter-term agendas. Many are primarily concerned with using the process to *appear* concerned, to *appear* to take action, and thus to obtain positive public relations and political cover, while avoiding, or delaying as long a possible, any real action. Given the nature of the IPCC review process, the authors of Chapter 2 would probably have found it impossible to raise questions about the motivations that underlie the FCCC or about how seriously it, and the activities it has sparked, should be viewed as decisionmaking processes designed to reduce the impacts of anthropogenic climate change.

Of course, the fact that participating governments have mixed motives does not mean that the FCCC will be of no consequence. The gradual accumulation of international accords and infrastructure could ultimately lead to a process that modifies significantly climate-related behaviors. On the other hand, to pretend that in its present form the FCCC is anything but a symbolic shell can lead us seriously astray. Talk of strategies such as joint implementation—by countries that are unwilling to impose even very small taxes on gasoline and other fossil fuel consumption—should raise serious doubts about what "problem" "decisionmakers" are actually addressing.

In its discussion of uncertainty, Chapter 2 differentiates between "scientific," "sociological," and "socioeconomic" uncertainty (IPCC 1996, 59). Such a classification can certainly be made, but I do not find it particularly useful because it provides no guidance on how the uncertainty originates, or how it should be dealt with in analysis. A more useful classification in these contexts would differentiate between circumstances in which

- the functional relationships among variables and the values of important coefficients are well known, but the initial conditions for the system are uncertain;
- coefficient values are uncertain or, while known for the moment, are evolving over time in an uncertain manner; and
- functional relationships among important variables are unknown or are evolving over time in an uncertain manner.

Most climate-change problems, whether they involve physical, biological, or social systems, involve all three of these types of uncertainty. Techniques for dealing with the first two types are well developed (Morgan and Henrion 1990) and have been applied in a number of analyses of the climate problem. Much less attention has been devoted to the development and practical demonstration of methods for dealing with uncertainties in model structure or in the evolution of that structure over time.

A few examples of analysis of the implications of model uncertainty do now exist in the climate literature. For example, Dowlatabadi has run a number of different versions of the ICAM-2 integrated assessment simulation model in order to explore how the conclusions reached depend upon the model structure that is assumed (Morgan and Dowlatabadi 1996). Table 3 provides an illustration. In this case five different versions of the ICAM-2 model were used to answer the question: "What is the probability that a $4.00/ton carbon tax that begins in the year 2000 and increases by $4.00/ton every five years through the year 2100 will have a positive net present value?" The different model forms that were assumed are listed in the left-hand column. Other cells in the table display the probability that the specific tax policy will have a positive net present value given the specific model assumptions. Model 1 (top row) applies the same discounting strategy everywhere, assumes autonomous technical change, does not include the radiative effects of aerosols, assumes that people will not adapt to impacts, and assumes that oil and gas reserves are exhausted by 2050. In contrast, Model 5 (bottom row) applies different discounting strategies depending on the different levels of economic development in different regions after Schelling (1994), assumes that carbon taxes induce accelerated technical change, includes the radiative effects of aerosols, assumes a high level of adaptation to impacts, and assumes that oil and gas reserves are not extinguished. Note that the probabilities which result range widely depending upon the structure of the model used to answer the question, as well as the geographical region that is considered.

A closely related issue, which I believe has also received far too little attention, is the question of how to appropriately perform analysis and display results when the problem involves components whose uncertainties evolve at different rates over time. Here a differentiation among physical, biological, and social systems may prove useful. Any analysis which combines models of socioeconomic process (whose uncertainty structure may evolve considerably over time scales of decades) with geophysical models (whose uncertainty structure often evolves more gradually) faces this problem. To date, I am not aware of any satisfactory treatments of these issues.

SPECIFIC CONCLUSIONS ABOUT DECISION ANALYSIS

I turn now to a consideration of the specific conclusions reached in Chapter 2. Five relate to decision analysis and five to negotiation. I will quote each in its entirety and then offer comments.

The first conclusion about decision analysis reads:

Table 3. Probability that Different Policies Will Have Positive Net Present Value as Determined in ICAM-2.

Structure of the model used to evaluate the policy	China	E Europe & former Soviet Union	India & SE Asia	Africa	Middle East	OECD nations	Latin America	Entire world
						Probabilities [a]		
Discount rate same in all regions; technology change is autonomous; radiative effect of aerosols is not included; impacts are permanent, no adaptation occurs; and no new resources of oil and gas are discovered	0.05	0.20	0.40	0.40	0.45	0.50	0.60	0.25
Discount rate depends on per capita growth in region; technology change is autonomous; radiative effect of aerosols is not included; impacts are permanent, no adaptation occurs; and no new resources of oil and gas are discovered	0.10	0.65	0.40	0.50	0.50	0.85	0.70	0.70
Discount rate depends on per capita growth in region; technology change is induced by carbon tax; radiative effect of aerosols is not included; adaptation occurs after impacts are detected; and new resources of oil and gas are discovered	0.10	0.70	0.55	0.60	0.55	1.00	0.95	0.95
Discount rate depends on per capita growth in the region; technology change is induced by carbon tax; radiative effect of aerosols is included; impacts are permanent, no adaptation occurs; and no new resources of oil and gas are discovered	0.05	0.40	0.30	0.30	0.40	0.60	0.70	0.55
Discount rate depends on per capita growth in region; technology change is induced by carbon tax; radiative effect of aerosols is included; adaptation occurs after impacts are detected; and no new resources of oil and gas are discovered	0.05	0.15	0.25	0.25	0.20	0.60	0.70	0.45
Discount rate depends on per capita growth in region; technology change is induced by carbon tax; radiative effect of aerosols is included; adaptation occurs after impacts are detected; and new resources of oil and gas are discovered	0.00	0.05	0.15	0.20	0.15	0.30	0.25	0.15

Note: The table illustrates, with results from the ICAM-2 integrated assessment model, that the answer which analysis provides to key policy questions can depend critically upon the structure of the model that is used.

a. These probabilities are those that the carbon tax policy described in the text will have a positive net present value in different geographical regions.

Source: Adapted from Morgan and Dowlatabadi 1996.

There is no single decisionmaker in climate change. Because of differences in values and objectives, parties participating in a collective decision-making process do not apply the same criteria to the choice of alternatives. Consequently, decision analysis cannot yield a universally preferred solution (IPCC 1996, 57).

The first two sentences in this conclusion are clearly true. However, the conclusion about decision analysis is not strictly true. Agreement is not required on values or on decision rules, only on actions. One can do separate analyses for different parties, applying their different values and decision rules. Sometimes, for very different reasons, the parties will all choose to take the same action. Figure 1 provides an example, based on work by Dowlatabadi using the ICAM-2 integrated assessment simulation model. In this case, different regional decisionmakers reach the same conclusion about what policy to adopt based on very different decision rules (Morgan and Dowlatabadi 1996).

Outcomes which improve (or do not worsen) the positions of all parties are termed Pareto improvements. Using a model called PARETO, Manne (1996) has recently illustrated such strategies in the context of the climate problem.

Chapter 2's second conclusion about decision analysis reads:

Decision analysis requires a complete and consistent utility valuation of decision outcomes. In climate change, many decision outcomes are difficult to value and a global welfare function does not exist, so quantitative comparisons of decision options are not meaningful (IPCC 1996, 57).

In this case I agree fully with the conclusion that "many decision outcomes are difficult to value and a global welfare function does not exist" but have problems with the balance of the conclusion. For real problems it is virtually *never* possible to specify all uncertainties and all outcomes, or to value all potentially relevant attributes in a utility function. Thus, virtually every application of decision analysis to real-world problems involves a *conditional* analysis. The analysts do the best job they can. Implicit in the results is a *ceteris paribus* statement across the variety of issues that it has not been possible to formally include. This is one of the principal reasons that most decision analysts recommend that the results of their analysis be used as guidance and not as the sole basis for decision-making. There may be important factors that it has not been possible to value and include in the analysis, which are relevant to the decision. The situation may be more extreme in the case of climate change, but in principle the same difficulty applies in most applications of decision analysis.

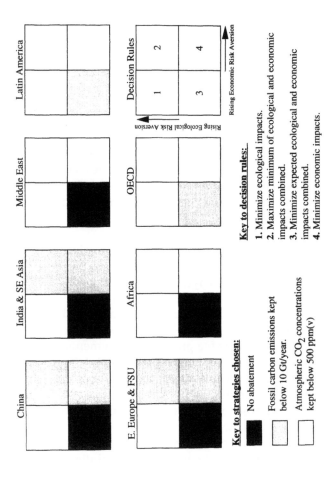

Figure 1. Agreement among Different Regional Decisionmakers on a Particular Climate Policy Does Not Require Agreement about the Decision Rules that Will Be Used.

Note: In this figure, based upon analysis performed using the ICAM-2 model, four rather different decision rules have been employed (denoted by the four quadrants of the boxes) to examine the policy choices (denoted by the shadings) that would be made by regional decisionmakers in each of seven different regions. Note that the same policy choice (same shading) can occur as a result of applying different decision rules (different quadrants) in seven regions. Details about the specific assumptions of the model and the specific formulation of the decision rules are not important to the basic point being illustrated.

Source: Reproduced from Morgan and Dowlatabadi 1996.

The third conclusion reads:

> Decision analysis may help keep the information content of the climate change problem within the cognitive limits of decisionmakers. Without the structure of decision analysis, climate change information becomes cognitively unmanageable, limiting the ability of decisionmakers to analyze the outcomes of alternative actions rationally (IPCC 1996, 57).

I agree that it is necessary to impose *some* structure on the climate problem to help make it cognitively manageable. However, classical decision analysis is not the only, and probably not the best, vehicle that can be used to impose such structure.

The fourth conclusion is that

> The treatment of uncertainty in decision analysis is quite powerful, but the probabilities of uncertain decision outcomes must be quantifiable. In climate change, objective probabilities have not been established for many outcomes, and subjective probabilities would be controversial, so climate change decisions cannot fully satisfy this requirement (IPCC 1996, 57).

I have never encountered a serious piece of decision analysis on a real policy problem that was based strictly upon "objective probabilities." By its very nature, decision analysis is a Bayesian undertaking, which means that many of the key uncertainties must be quantified with subjective judgment. The uncertainties may be larger, and their nature and structure more complex in the case of climate, but I see no fundamental obstacle to the development and use of subjective probability judgments. As noted above, Keith and I have recently completed a program of detailed expert elicitation of a group of sixteen climate scientists[1] (Morgan and Keith 1995). Figure 2 and Table 4 summarize a few of the results. Nordhaus (1994) has completed a similar set of interviews involving economic judgments about market and nonmarket impacts.

There is, of course, no agreement among experts about the probabilities that should be used to describe the various uncertainties associated with the climate problem. In such circumstances, a standard technique in decision analysis—when applied to public-sector problems—is to perform a comparative analysis, examining the extent to which the disagreements among the experts are important in the context of the policy questions being examined. In some cases, they may matter a great deal, in others, they may be entirely irrelevant.

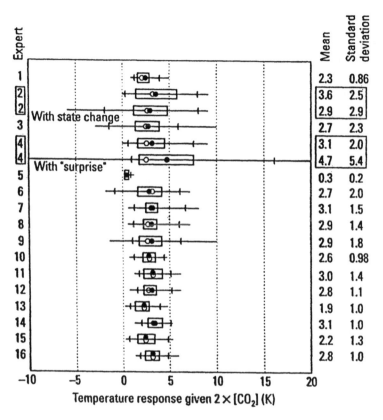

Figure 2. Estimates by Sixteen Leading U.S. Climate Scientists of the Amount of Warming that Would Result from Doubling the Concentration of Atmospheric CO_2.

Note: Horizontal lines indicate the range from the minimum to the maximum assessed possible values. Vertical tick marks denote the lower fifth and upper ninety-fifth percentiles. Boxes indicate the interval spanned by the 50% confidence interval. Solid dots are the mean and open dots the median estimates.

Source: Reproduced from Morgan and Keith 1995.

The final conclusion reached in Chapter 2 in connection with decision analysis reads:

Because of the large uncertainties and differences between parties, there may be no "globally" optimal climate change strategy; nevertheless the factors that affect optimal single-decision-maker strategies still have relevance to individual parties (IPCC 1996, 57).

Table 4. Uncertainty in Present Scientific Understanding of Global Climate Factors.

Factor contributing uncertainty	Number of experts assigning rank of:					Mean uncertainty[a]
	1	2	3	4	5	
Cloud distribution and optical properties	9	1	0	1	3	38%
Convection/water vapor feedback	6	3	2	0	0	35%
Carbon dioxide exchange with terrestrial biota	1	1	3	5	1	19%
Carbon dioxide exchange with the oceans	0	1	5	2	2	16%
Oceanic convection	0	3	0	1	1	22%

Notes: Uncertainty in the present scientific understanding of the five factors listed in the left-most column was most frequently cited by sixteen leading U.S. climate scientists as the largest contributor to their overall uncertainty (displayed in Figure 2) about the amount of warming that would occur if the concentration of atmospheric CO_2 were doubled. The numbers in the center columns display the number of experts who ranked each factor as being the first through fifth most important source of uncertainty. The right-most column indicates the mean of the experts' estimates of how much uncertainty would be reduced overall if all uncertainty associated with this factor could be eliminated.

a. The percentages in this column are the mean uncertainty reductions in the amount of warming if all uncertainty in each of these factors were eliminated.

Source: Based on Morgan and Keith 1995

While I agree with the basic idea, I would be considerably more comfortable with a more conditioned statement such as, "Because of the large uncertainties and differences between parties, no single climate change strategy is likely to be optimal for everyone; nevertheless many of the factors that affect optimal single-decision-maker strategies can be expected to have relevance for many of these parties."

SPECIFIC CONCLUSIONS ABOUT NEGOTIATION

We turn now to the five conclusions reached in Chapter 2 on the topic of negotiation. The first reads:

> Excessive knowledge requirements in negotiated environmental decisions may stand in the way of collective rational choice. This difficulty could be reduced by making the negotiation process itself more manageable through the use of tools like stakeholder analysis or by splitting accords into more easily managed clusters of agreements (IPCC 1996, 57).

I agree that various forms of analysis may help facilitate negotiation processes, although eliminating uncertainty does not always facilitate negotiation (Raiffa 1982). Sometimes uncertainty and ambiguity provide the space, and the political cover, that different parties need if they are to reach agreement on a single course of action.

I am concerned about the extent to which an objective of analysis should be to simplify the problem. In a number of recent discussions of the negotiating process, I have heard the argument that the issues must be *dramatically* simplified in order to make international negotiations possible. While we obviously should strive for simplicity, I believe that the greater risk is oversimplifying issues to the point that resulting agreements do not adequately address the problems.

It is not true that international agreements are only possible in highly simplified environments. Commercial agreements, negotiated internationally among firms, often involve highly complex technical issues. Agreements between governments on international technical standards, in fields such as communications, aviation, and taxation, are often similarly complex. In negotiations among nations, nuclear arms control provides an excellent example of a series of agreements reached in the context of a highly complicated technical and political setting. These were not achieved by dramatic simplification which allowed diplomatic generalists to handle the problem. Rather, groups of expert negotiators have spent their entire careers mastering the relevant technical, military, and diplomatic complexities of the problems. There is no reason why we should not demand the same level of specialized expertise from negotiators who address issues of climate change.

The concept of "Global Warming Potential" clearly illustrates the risks associated with oversimplification.[2] Many diplomats frame the climate problem as one of negotiating international accords that specify how countries will manage their greenhouse gas emissions over time. To simplify the problem of specifying what nations should and have done, they have sought a single metric that can be used to compare emissions across nations and over time. The most widely discussed metric is the Global Warming Potential, or GWP (Lashof and Ahuja 1990), which compares the net radiative forcing of a unit of any other trace gas (such as methane), over its lifetime as gas, with that of a unit of carbon dioxide.

Different human activities produce different mixes of greenhouse gases. For example, electricity generation mainly produces carbon dioxide. In contrast, rice farming mainly produces methane, which has an instantaneous radiative forcing an order of magnitude larger than carbon dioxide but an atmospheric residence time an order of magnitude shorter.

GWPs capture the instantaneous physics of the problem. By integrating for some arbitrary period, the length of which is a value judgment and thus subject to controversy, GWPs can also reflect some aspects of the temporal dynamics. However, they do not recognize the economic reality of discounting, which can work to reduce the effective importance of future changes in radiative forcing (Eckaus 1992). Further, Kandlikar (1994) has argued that, because most decisionmakers are principally concerned about impacts, not gas concentrations, any reliable equivalence scheme also needs to incorporate climate change and impact dynamics.

Kandlikar has proposed that the problem can be addressed by devising indices that compare the eventual economic impacts of unit emissions of different greenhouse gases (Kandlikar 1995). Hammitt and co-workers have more recently advanced similar ideas (Hammitt and others 1996).

If greenhouse damages are a function of global mean temperature change, then the indices will depend on the future emissions of trace gases. Future emissions of trace gases are intrinsically linked to economic growth and abatement policies, which in turn are governed by expectations of greenhouse damages. Trace gas indices can thus be calculated either on the basis of emissions scenarios, such as those devised by the IPCC, or using optimal control techniques where the trade-off between damages and abatement costs is made explicit, and the trace gas index values are a by-product of computing an optimal emissions trajectory.

Damages and abatement costs are poorly known quantities. However, by using a range of plausible values for these quantities it is possible to provide values for a scenario-based index, or SBI, and draw some conclusions about their dependence on uncertain variables. Results are shown in Table 5. While for long-lived nitrous oxide, the difference between the GWP and the SBI is 10% or less, the difference can be much greater for species with atmospheric lifetimes significantly shorter than CO_2. The differences exceed a factor of three for methane and a factor of two for HCFC-22. Such differences can have *enormous* economic consequences in the abatement costs that a country, especially one with a large agricultural sector, may face.

The second conclusion related to negotiation reads:

> Since society has no consistent probability threshold for ignoring particular risks, it may be vulnerable to surprise when risks are uncertain. In climate decisions, this vulnerability could be reduced by relating event scenarios to explicit probabilities of surprise (IPCC 1996, 57).

This is far too tentative a conclusion. Both individuals and society *are* often vulnerable to surprise from low probability events, even when the probabilities involved are well known, as they frequently are in the case of hurricanes and floods. This occurs for several reasons: as a result of the heuristic procedures that people use when reasoning about uncertain events (Kahneman and others 1982; Dawes 1988); because of an unwill-

Table 5. Comparison of Several Different Indices for Radiatively Important Trace Gases.

			SBI: Square[c]		*SBI: Cube[d]*	
Trace gas	*GWP[a]*	*SBI: Linear[b]*	*IPCC-A*	*IPCC-D*	*IPCC-A*	*IPCC-D*
Methane						
2% discount	11	19	12	12.9	8.5	10
6% discount	11	38.4	27.5	28.1	19.9	21.5
Nitrous Oxide						
2% discount	290	269	282	280	289	286
6% discount	290	258	271	269	278	275
HCFC-22						
2% discount	1500	2445	1706	1811	1284	1466
6% discount	1500	3178	2217	2354	1700	1879

Notes: This table compares global warming potential (GWP) values with several alternative scenario-based indices (SBI) for trace gases. All values shown compare other trace gases with carbon dioxide. Thus, an index value of 100 means a unit mass of the trace gas contributes 100 times as much as a unit mass of carbon dioxide. IPCC-A and IPCC-D refer to two standard emissions scenarios in the IPCC's 1990 assessment.

a. GWP computed in this example with a time horizon of 100 years.

b. SBI index when damage is assumed to be linearly proportional to temperature. (If the damage is linear in temperature, details of the emissions scenario do not affect the relative weights for the trace gases.)

c. SBI index when damage is assumed to be proportional to the square of the temperature.

d. SBI index when damage is assumed to be proportional to the cube of the temperature.

Source: Reproduced from Morgan and Dowlatabadi 1996.

ingness or inability to think seriously about the future; and as a result of the impact of short-term interests and interest group politics.

In decision analytic terms, the choice to ignore particular risks should be based on a joint consideration of the probability involved and the consequences. Individuals and societies are certainly not consistent about where they set the *de minimus* level of risks. Risk ranking and other decision aids now being developed (Davies 1996) may help individuals and societies be more consistent in their choices, and hence more efficient and equitable in their risk-management decisionmaking. However, I do not believe that this inconsistency in choosing the *de minimus* level of risk is the principal cause of vulnerability to surprise.

The third conclusion about negotiation is that

> In the face of long-term uncertainties, sequential decisionmaking allows actions to be better matched to outcomes by incorporating additional information over time. Sequential decisionmaking also minimizes harmful strategic behavior among multiple decision-makers (IPCC 1996, 57).

The recognition that decisionmaking about climate change will be sequential is important and has been absent in too many analyses of this subject. The discussion in Chapter 2 reads as if the world has some choice about whether to adopt sequential decisionmaking. In fact, whatever the desirable or undesirable traits of sequential decisionmaking, the world has very little choice but to adopt a sequential approach. While we might agree on a set of decisions today, we cannot completely bind later generations to abide by those decisions. Social and political reality dictates that choices about climate change will inevitably be made through multiple, parallel processes of sequential decisionmaking. Whether this will allow us to make effective use of evolving understanding depends on the relative time constants of global-change processes and human institutions and technologies.

The chapter includes a discussion of "act-learn-act" sequential strategies, which are clearly important. One key point that is not discussed is that not all actions provide equal opportunity to learn. If the world is truly going to try to adopt an "act-learn-act" strategy, and not simply use it as an excuse to delay difficult and costly choices, the actions chosen must be carefully designed to maximize the opportunities to draw useful and correct lessons.

The authors should be commended for stressing five types of short-term action which can considerably improve the option set for future decisionmakers. These are climate research, basic technology research (although it is unfortunate that the chapter addresses only government-led programs), market incentives to induce technical change, adoption of low-cost abatement options today, and the exercise of care to avoid choices today that might lock in carbon-intensive future development paths.

I am unpersuaded by the assertion that sequential decisionmaking "minimizes harmful strategic behavior among multiple decisionmakers" (IPCC 1996, 57). The literature on organizational behavior (March and Olsen 1979) suggests quite the opposite. In Kingdon's language, sequential decisions involve the creation of a series of policy windows. Kingdon writes, "An open window is an opportunity for advocates to push their pet solutions or to push attention to their special problems. Indeed, advocates in and around government keep their proposals and their problems at hand, waiting for these opportunities to occur. They have pet solutions, for instance, and wait for problems to float by to which they can attach their solutions, or for developments in the political stream that they can use to their advantage. Or they wait for similar opportunities to bring their special problems to the fore..." (Kingdon 1984, 212–13). This hardly sounds to me like a way to minimize "harmful strategic behavior."

Because of sequential decisionmaking in a political process that involves multiple actors with multiple agendas, sustaining a coherent policy over time can be very difficult. The design evolution of large government-managed systems that have not enjoyed the political cover of "defense" or the "Cold War," such as the space shuttle, space station, Clinch River Breeder Reactor, and Super Conducting Super Collider, suggests to me that certain kinds of technical systems may not be compatible with the social processes of sequential decisionmaking. The conclusion I draw is not that we should try to change social reality, but that we should recognize the constraints imposed by that reality and choose technical solutions that are likely to be compatible with those constraints.

The chapter argues that "the IPCC negotiations do not have to resolve controversies on long-term issues.... The objective of the first step of the decision process is to put society as far as possible in the position to postpone technological or institutional 'lock-in' and use the extra negotiation time to increase options and reach wider consensus on how to approach the more difficult longer-term decisions" (IPCC 1996, 71). As noted above, the chapter offers a set of prudent policy prescriptions. However, I believe that the discussion that leads to these recommendations is framed in an unrealistically optimistic way. We already have very substantial "technological and institutional lock-in" (71). We need policies such as "low-cost abatement decisions to increase learning" (71) in part to begin to counter the existence of lock-in.

While we may succeed in reducing somewhat the magnitude of climate change and its impacts, the very long geophysical and socioeconomic time constants involved suggest that it is extremely unlikely that significant change can be avoided. Given social and technical time constants, it seems likely to me that the longer we wait before undertaking substantial reductions in emissions, the greater the changes will be. Political reality dictates that we have no choice but to engage in a process of sequential decisionmaking...but let's not go overboard in trying to make a virtue out of this necessity.

The next conclusion is that

> Improved information about uncertain outcomes may have very high economic value, especially if that information can create future decision options (IPCC 1996, 57).

I agree. However, I am left a bit uncomfortable with two aspects of the discussion associated with this conclusion in the body of the chapter. First, the traditional definition of the value of information is framed in terms of a *specific* decision. Second, uncertainty is assumed to decrease when resources and time are invested in research. Neither of these assumptions is safe in the context of research on an issue in which scientific knowledge is still very incomplete, such as the climate problem. Given the considerable uncertainty about what the right decision problems are, one should be careful about adopting any specific decision analytic framing to compute the value of information. Information that appears to have little or no value in one framing of the problems may prove to be of considerable value in other framings. Further, in a problem such as climate change and its impacts, research often has the consequence that it *increases* uncertainty for a considerable period before it ultimately leads to a reduction. In the expert elicitation Keith and I performed (Morgan and Keith 1995), we asked the sixteen climate experts to allocate a budget of $1 billion per year for fifteen years in such a way as to yield the greatest likelihood of reducing uncertainty about climate sensitivity and a number of similar quantities. We later asked a set of questions about surprise, which included a judgment of the probability that, after an expenditure of $15 billion and the passage of twenty years, the experts' judgments about the uncertainty associated with the value of climate sensitivity would increase by more than 25% over their level today. The results, reported in Table 6, show that different experts judge that there is between an 8% and a 40% chance they would be *less* sure (by >25%) in their estimate of climate sensitivity

Table 6. Estimates of How Much Climate-Change Research Will Reduce Uncertainty.

Expert number	Chance of uncertainty[a]
1	10
2	18
3	30
4	22
5	30
6	14
7	20
8	25
9	12
10	20
11	40
12	16
13	12
14	18
15	14
16	8

Notes: Estimates made by sixteen leading U.S. climate scientists indicate that, after $15 billion of optimally designed research and twenty years, their uncertainty about how much warming would result from a doubling of atmospheric concentration of CO_2 would be more than 25% larger than it is now.

a. Chance that uncertainty rises more than 25% after twenty years of research.

b. Expert 3 used a different response mode for this question, he gave a 30% chance of an increase by a factor of $2.5.

Source: Based upon Morgan and Keith 1995.

after this program of research than they are today. This comes as no surprise to most scientists but is a possibility that many decisionmakers, and even some decision analysts, tend to overlook.

Clearly there is a value to learning that you knew less than you thought you knew. In the context of any specific decision, this value can be monetized. However, when the consequence is that you conclude that you have framed the analysis incorrectly, and should really be considering a different decision, monetizing the value of the information becomes rather more difficult.

The chapter's final conclusion about negotiation reads:

> There are currently no effective mechanisms for the sharing of risks related to climate change and their associated economic burdens. International risk sharing could yield substantial benefits for global economic and social welfare (IPCC 1996, 58).

Within a nation, particularly a developed democracy, a number of mechanisms exist for sharing such risks, including private insurance, state-subsidized insurance, public and private disaster relief, capital markets, manpower retraining programs, extension services, and investment in infrastructure. At the international level, the institutions are weaker. While certainly not absent, current mechanisms for risk sharing between the developed and the undeveloped world are weak and unreliable.

Beyond noting the need for more effective risk-sharing mechanisms, I would offer three warnings. First, there are cultures in which social risk sharing as suggested here is not a well-established concept. Often these societies are fatalistic, ascribing all outcomes to fate or an omnipotent God. Second, as I suggested above, there is a risk that developed countries such as the United States will try to use international programs, such as joint implementation, as political cover to avoid taking even modest domestic steps such as imposing emissions taxes. Third, many arguments about emissions trading and joint implementation ignore the fact that many developing countries are supply constrained with respect to energy. They also ignore factors, such as corruption, which could work to exaggerate effective national discount rates.

LIMITS TO THE CONVENTIONAL TOOLS FOR POLICY ANALYSIS

As policy analysts have turned their attention to the problem of climate change, they have found it natural to employ the conventional tools of quantitative policy analysis, such as utility theory, benefit-cost analysis, statistical decision theory, multi-attribute utility theory, contingent valuation, and a variety of similar methods. One encounters descriptions of such applications throughout the report of Working Group III.

These familiar tools were developed for, and have typically been applied to, problems within a specific political and cultural setting. Most often they support decisionmaking within a relatively homogeneous cultural and political setting that involves socially modest expenditures on time scales comparable to, or shorter than, the lifetimes of the players.

To illustrate the point, consider the three dimensional space shown in Figure 3. The vertical axis shows time. In locating a policy action along

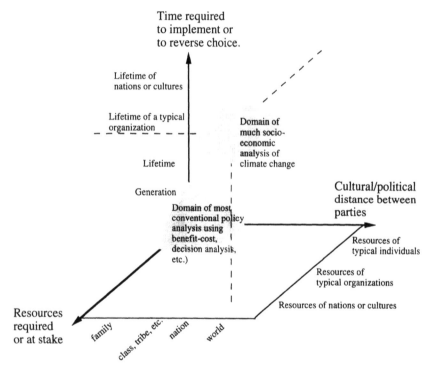

Figure 3. A Useful Space for Classifying Policy Analytic Tools.

Note: Policies, and the actions needed to undo their effects, can be located in this space. For any particular policy problem the axes indicate the amount of resources required or at stake; the time scale—both to implement and to reverse the effects of the choices available; and the degree of political and cultural homogeneity of the people involved. Most tools of modern policy analysis were developed to address problems that lie near the origin in this space. As one moves outward from the origin, more and more of the underlying assumptions on which these tools are based require careful examination. Many problems of global change lie far from the origin on all three axes. Global change issues may involve very large costs, are often characterized by long temporal scales and associated intergenerational equity issues, and may involve a large political and cultural distance between many different parties and an associated lack of shared metrics.

Source: Adapted from Morgan and others 1996.

this axis, it is important to consider both the time required to implement the policy as well as the time required to undo the consequences of that policy. For some policies the two values are very similar. For others they can lie very far apart.

The resource axis ranges from the level commanded by typical individuals to the level that entire nations or cultures can command. Note that the level of resources required to implement a policy may be quite different from the level needed to undo its effects. Sometimes the cost of undoing things is much higher than that of the original action. For example, it may be quite cheap to release a pollutant into the environment and extremely expensive to remove it once it has been widely dispersed within an ecosystem. Sometimes the reverse is true. For example, burning down a building is typically far cheaper than building it.

The axis labeled "cultural/political distance between parties" refers to the parties involved in, as well as those that are significantly affected by, the policy choices. In most applications of policy analysis, the distance is relatively short. For example, the participants may all be middle-class Americans.

The conventional tools of quantitative policy analysis were primarily developed for, and have primarily been applied to, policy problems that lie close to the origin in Figure 3. When they are applied to problems that lie farther out from the origin—such as a decision to clear-cut old growth redwood forests (which requires few resources and little time to implement but thousands of years to undo the consequences) or a decision to place a strip mine at Black Mesa, Arizona (which involves a very great cultural distance between Anglo-American power officials and Navaho religious leaders)—they tend not to work reliably or they produce results which some players find unacceptable.

Many problems in climate policy involve temporal, spatial, and sociopolitical scales that are significantly broader than those encountered in traditional policy analysis. That is, important aspects of climate problems often lie far out from the origin on all three axes in Figure 3. Because of this, one should be cautious in applying conventional tools of policy analysis to these problems.

Such tools frequently assume the existence of a single decisionmaker and are often based on maximizing that decisionmaker's utility. Many are designed for addressing impacts which are of manageable size and valued at the margin. Most assume the existence of values that are exogenously determined and static over the lifetime of the policy. Some implicitly assume that there is a single polity with a legal structure which can enforce rules of conduct and the terms of contractual agreements. Most employ conventional discounting to combine or to compare costs and benefits that occur at different moments in time. While some tools, most notably decision analysis, are designed to deal with uncertainty, the level of the uncertainty is typically assumed to be manageable (model structure known, coefficient values uncertain across some modest range, structure of the uncertainty stationary over time). Finally, in order to simplify the analysis, much conventional policy analysis assumes local system linearity.

Each of these assumptions is violated by some parts of the climate problem. In other writing, my colleagues and I have begun to explore such issues (Morgan and Dowlatabadi 1996; Morgan and others 1996). In another paper in this volume, Robert Lind and Richard Schuler explore a number of the problems associated with discounting.

For the purposes of this paper it is sufficient to note that there is a potential problem. When the next Working Group III produces the next review of the economic and social dimensions of the climate problem, I believe that it should devote much more attention to examining the appropriateness of the analytical tools that have been employed in the literature. Such a review could be useful to the community of researchers working on climate impacts and policy and might also help make the next product of Working Group III more acceptable to those of us who are uncomfortable with some of the results obtained by applying conventional policy analysis tools to climate problems.

IMPROVING THE LITERATURE BASE
FOR THE NEXT IPCC REPORT

I began this paper by noting that the authors of Working Group III faced the disadvantage of a relatively sparse literature on analytical methods specifically related to climate change. If the situation is to be any better when the next IPCC review occurs, two things must happen. A number of thoughtful researchers working on climate issues need to think critically about the strengths and limitations of alternative analytical tools and strategies in the specific context of the climate problem and publish their arguments and conclusions in the literature. Second, and even more important, an expanded research effort must be mounted on methods development and evaluation.

I see hopeful signs on both accounts. Several chapters in the Working Group III report prompted vigorous debate in the climate community and appear to have stimulated a number of investigators to devote more long-term research attention to issues of methods. In addition, as new researchers from various disciplines have begun to work on climate, they have brought with them additional insights. For example, in the process of creating the National Science Foundation–supported Center for the Integrated Study of the Human Dimensions of Global Change at Carnegie Mellon, several experts on the behavioral-experimental literature on time preferences have become aware of climate issues and begun to think about what that literature can say to inform researchers working on climate problems.

There are also signs for optimism in the area of new research. In the United States, the National Science Foundation Program on Methods and Models in Integrated Assessment has begun to support research on evaluating and improving existing analytical methods and developing new methods. This is a particularly important development because, historically, support for policy-focused methods-related research has been extremely difficult to find.

From an analytical perspective, climate change is a very tough problem. Even with expanded interest and research support, progress may be slow. But any progress we can make will be important, both from the perspective of improving the advice and insights that can be provided to private and public sector decisionmakers, and from the perspective of setting a new benchmark for how to do social and policy analysis on problems such as climate in which complex issues of science and technology are of central importance.

ENDNOTES

1. Together with Louis Pitelka, I am currently planning a similar set of detailed interviews with experts on terrestrial ecosystem impacts.

2. The discussion is drawn from Morgan and Dowlatabadi (1996).

ACKNOWLEDGMENTS

I thank Hadi Dowlatabadi, Paul Fischbeck, Baruch Fischhoff, Milind Kandlikar, David Keith, Lester Lave, Anand Patwardhan and James Risbey for ideas and helpful conversations. The work was supported by the Electric Power Research Institute (RP-3441-14), the National Science Foundation (SES-9022738, SBR-9209783, SBR-9521914, DMS-9523602), the Department of Energy (DE-FG02-94ER61916), and Carnegie Mellon University.

REFERENCES

Allison, Graham T. 1971. *Essence of Decision: Explaining the Cuban Missile Crisis.* Boston: Little Brown and Co.

Davies, J. C., ed. 1996. *Comparing Environmental Risks: Tools for Setting Government Priorities.* Washington, D.C.: Resources for the Future.

Dawes, R. M. 1988. *Rational Choice in an Uncertain World.* Harcourt, Brace and Jovanovich.

Eckaus, R. 1992. Comparing the Effects of Greenhouse Gas Emissions on Global Warming. *Energy Journal* 13(1): 25–35.

Hammitt, J. K., A. K. Jain, J. L. Adams, and D. J. Wuebbles. 1996. A Welfare-Based Index for Assessing Environmental Effects of Greenhouse-Gas Emissions. *Nature* 381:301–3.

IPCC (Intergovernmental Panel on Climate Change). 1996. *Climate Change 1995: Economic and Social Dimensions of Climate Change.* Edited by J. P. Bruce, H. Lee, and E. F. Haites. Cambridge: Cambridge University Press.

Kahneman, D., P. Slovic, and A. Tversky, eds. 1982. *Judgment Under Uncertainty: Heuristics and Biases.* Cambridge University Press.

Kandlikar, M. 1995. On the Relative Role of Greenhouse Gases in Abatement Policies. *Energy Policy* 23(10): 879–83.

———. 1994. *Reconciling Uncertainties in Integrated Science and Policy Models: Applications to Global Climate Change.* Ph.D. thesis, Carnegie Mellon University.

Kingdon, John W. 1984. *Agendas, Alternatives and Public Policies.* Boston: Little Brown and Co.

Lashof, D. and D. Ahuja. 1990. Relative Contributions of Greenhouse Gas Emissions to Global Warming. *Nature* 344:529–31.

Manne, A. S. 1996. Greenhouse Gas Abatement: Toward Pareto-Optimality in Integrated Assessments. In *Education in a Research University,* edited by K. J. Arrow, R. W. Cottle, B. C. Eaves, and I. Olkin. Palo Alto: Stanford University Press.

March, J. G. and J. P. Olsen. 1979. *Ambiguity and Choice in Organizations.* Oslo: Universitetsforlaget.

Morgan, M. G. and M. Henrion. 1990. *Uncertainty: A Guide to Dealing with Uncertainty in Quantitative Risk and Policy Analysis.* New York: Cambridge University Press.

Morgan, M. G. and D. Keith. 1995. Subjective Judgments by Climate Experts. *Environmental Science and Technology* 29(10): 468A–76A.

Morgan, M. G. and H. Dowlatabadi. 1996. Learning from Integrated Assessments of Climate Change. *Climatic Change* 34:337–68.

Morgan, M. G., M. Kandlikar, J. Risbey, and H. Dowlatabadi. 1996. *Why Conventional Tools for Policy Analysis Are Often Inadequate for Problems of Global Change.* Department of Engineering and Public Policy, Carnegie Mellon University.

Nordhaus, W. D. 1994. Expert Opinion on Climate Change. *American Scientist,* January/February: 45–51.

Raiffa, H. 1982. *The Art and Science of Negotiation.* Cambridge, Massachusetts: Belknap Press of Harvard University Press.

Schelling, T. C. 1994. Intergenerational Discounting. *Energy Policy* 23:395–402.

———. 1978. *Micromotives and Macrobehavior.* New York: W. W. Norton.

Comments

Policy Analysis for Decisionmaking about Climate Change

Akihiro Amano

In his paper, Policy Analysis for Decisionmaking about Climate Change, M. Granger Morgan discusses questions concerning decisionmaking in the global community with respect to climate change in the Working Group III section of the IPCC Secondary Assessment Report. He also raises some specific issues related to Chapter 2 of the WGIII report. His discussion covers many important areas with both wider and narrower scopes, and I am sympathetic to many of his points, especially those dealing with distributional considerations, willingness-to-pay versus willingness-to-accept-compensation, discounting, cultural differences, and so on. However, in order to fulfill the duty of a discussant, I should like to present some comments that may reflect slightly different views, or alternative interpretations. I have organized my comments according to the order of Morgan's paper.

GENERAL CONCERNS

Morgan complains that Chapter 2 of the WGIII report largely adopts the perspective of the Framework Convention on Climate Change. It is true that decisions concerning anthropogenic causes of climate change involve millions of agents and that national governments may not have

AKIHIRO AMANO is professor of Economics at Kwansei Gakuin University, School of Policy Studies.

sufficient power to control them. On the other hand, members of the WGIII were subject to the work plan and terms of reference established by the IPCC Bureau. The work plan states, "The Working Group will, in accordance with its Terms of Reference, ...prepare a comprehensive technical assessment of the socio-economics of mitigation of climate change, the impacts of climate change and adaptation to climate change over both the short and long term and at the regional and global levels." Moreover, the review process of the report involves government reviews as well as expert reviews, and one of the major purposes of the report is to provide policymakers with useful information on climate-change policies. Therefore, it appears unrealistic to expect this report to manifest doubts about the ability of its clients (that is, governments of the parties). A Japanese saying may liken it to getting a fish from a tree.

Having said this, however, I would like to agree with Morgan that many of the approaches he cites in his Table 2, especially the prescriptive approaches, have not been dealt with in the WGIII report. Since the activities of the IPCC are expected to continue in the future, the evaluation of current research on these neglected areas, especially legal, political, and organizational models or theories, should receive sufficient attention within the IPCC framework.

From my standpoint, the most interesting contribution of Morgan's paper is the statement that "Much less attention has been devoted to the development and practical demonstration of methods for dealing with uncertainties in model structure, or in the evolution of that structure over time." Table 3 of his paper illustrates the point very clearly.

I have done a similar exercise using the MERGE model developed by Manne, Mendelsohn, and Richels (1995). I found that the choice of discount rates and endogenization of the costs of backstop technologies are the two most important factors affecting the choice of abatement timing. Figure 1 compares Scenarios A and B, where Scenario A is a Pareto-optimum scenario as described by Manne, Mendelsohn, and Richels (1995) and Scenario B assumes that the energy costs of backstop technologies are endogenously determined by a cumulative level of R&D spending in developed regions:

Cost of Backstop =

> Exogenously Given Costs $*\exp(-\mu * \text{Cumulative R\&D Investment})$

If economic agents know that abatement costs will decrease with the introduction of backstop technologies in the future, the assumption of exogenously introduced backstops has a tendency to delay emission-reduction decisions. Insofar as the costs of backstop technologies are

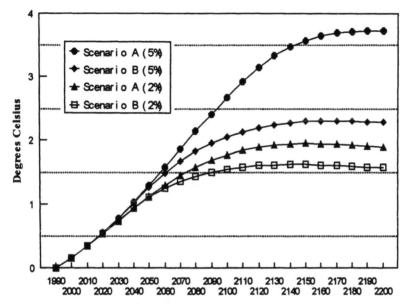

Figure 1. Timing of Emission Reduction.

affected by cumulative R&D expenditures, however, current consumption-savings decisions can influence the timing of abatement policies. Figure 1 clearly shows that incorporating this assumption introduces the tendency of intertemporal substitution of abatement efforts after the middle of the next century.

SPECIFIC CONCERNS

Decision Analysis

The question raised in the second point on page 34 of Morgan's paper concerns not so much the distinction between objective and subjective probabilities as the disagreements about the subjective probabilities. In this respect, the standard technique of comparative analysis suggested by the author should be used more extensively.

Negotiation

Morgan does not consider it appropriate to recommend attempts to change social reality. Rather, he advocates choosing technical solutions that are likely to be compatible with the constraints imposed by a political reality that we cannot change but which we can influence. The line of demarcation seems to be rather blurred, and its position may very much depend upon the scientific knowledge we possess and the institutional framework at the time of the decisions. Within the time frame of five to ten years or longer, much about these factors appears to be alterable.

As to international risk-sharing proposals related to climate change, Morgan offers three warnings. I would like to know his assessment of the liability schemes that have been put forth from somewhat different perspectives, such as the OASIS proposal. It seems to me that such a proposal is not susceptible to the dangers raised by the author. However, this question may be more appropriately relegated to the discussion of Chapter 3 of the scientific assessment review (SAR) (see IPCC 1996, 101–3).

CONCLUDING REMARKS

In the final section of his paper Morgan points out that the problem of climate change has many dimensions—such as the resources involved, the length of time lags, and cultural/political distances between parties—so we need to be cautious in applying the conventional framework of policy analyses to the present problem. Of course, the authors of the SAR must have been well aware of such extended fields, especially as far as the first two dimensions are concerned. In hindsight, the major problems I find are that the selection of lead authors and contributors was narrowly limited (almost) to the discipline of economics and the task of the writing teams was confined to the review of existing (published) literature. Future work should involve researchers from wider areas and invite them to make new contributions as well as review existing literature.

REFERENCES

IPCC (Intergovernmental Panel on Climate Change). 1996. *Climate Change 1995: Economic and Social Dimensions of Climate Change.* Edited by J. P. Bruce, H. Lee, and E. F. Haites. Cambridge: Cambridge University Press.

Manne, A. S. 1996. Greenhouse Gas Abatement: Toward Pareto-Optimality in Integrated Assessments. In *Education in a Research University*, edited by K. J. Arrow and others. Palo Alto: Stanford University Press.

Manne, A. S., R. Mendelsohn, and R. Richels. 1995. MERGE: A Model for Evaluating Regional and Global Effects of GHG Reduction Policies. *Energy Policy* 23(1):143–72.

Comments

Policy Analysis for Decisionmaking about Climate Change

Alan S. Manne

Many people believe that a horse is a more graceful animal than a camel. It is widely supposed that this is because the camel was originally intended to be a horse but was instead designed by a committee. By the same token, despite the talents of the individual authors of Working Group III's Chapter 2, their committee product is not highly readable. For those who are already familiar with economics, negotiation theory, and decision analysis, the WGIII Chapter 2 provides a handy and encyclopedic review of these ideas insofar as they relate to climate change. For nonspecialists, it is likely to be less helpful.

Perhaps this is not so surprising. The IPCC is too closely related to the United Nations to expect much more than a series of ingenious compromises between the views of strong-minded individuals. Successive drafts have been rewritten so as to incorporate almost everyone's ideas, and still meet a publication deadline. This volume will stand on the shelves of many political leaders but will be read by very few of them.

Let me begin by mentioning some of the good things that appear in Chapter 2. Each of the following ideas appears in the introduction on page 57:

> "...unique characteristics of climate change: large scientific and economic uncertainties; long time horizons; nonlinearities and irreversibility of effects; the global nature of the problem; social,

ALAN S. MANNE is Professor Emeritus of Operations Research at Stanford University.

economic, and geographic differences among the affected parties; and an agreed framework to address the issue."

"...in the face of long-term uncertainties, sequential decision
making allows actions to be better matched to outcomes by incorporating additional information over time."

"...decision analysis may help keep the information content of
the climate change problem within the cognitive limits of decision makers."

To my way of thinking, the principal sin of commission is the following paragraph. A careless reader could easily draw the conclusion that
there really is no role for quantitative analyses in this area.

Decision analysis requires a complete and consistent utility valuation of decision outcomes. In climate change, many decision outcomes are difficult to value and a global welfare function does
not exist, so quantitative comparisons of decision options are not
meaningful.

Later on page 57, the authors continue in this same vein: "The lack of
an individual decision maker, utility problems, and incomplete information suggest that decision analysis cannot serve as the primary basis for
international climate change decision making. Although elements of the
technique have considerable value in framing the decision problem and
identifying its critical features, decision analysis cannot identify globally
optimal choices for climate change abatement."

In effect, the authors have built a straw man. They describe a textbook form of decision analysis. They then discover that this approach is
unable to deal with all the practical problems of global climate change. If
the phrase "primary basis" had been replaced with "only basis," the conclusion would have been far more reasonable.

The authors alternate between the harsh realities of multilateral
negotiations (65–67) and the implicit approval of extreme risk avoidance
criteria such as a maximin criterion (63). It is, for example, suggested that
the international community might adopt a "safe minimum standard to
avoid dictatorship of the present" over future generations" (60). The latter
phrase is superficially attractive, but remarkably ambiguous. It will
appeal to those trained in the legal profession, and makes it all but impossible to analyze tradeoffs between conflicting objectives. It is the same
type of reasoning that once led the U.S. Congress to adopt the Delaney
approach to potential carcinogens in processed foods. That is, no matter

how low the carcinogen content—as long as it could be measured—the food item was forbidden.

As I read Chapter 2, here are the principal sins of omission:

In describing alternative decision frameworks on page 63, the authors concentrate on optimization as an approach for *individual* decisionmaking but fail to mention the criterion of Pareto-optimality. This leads them to neglect the possibilities for separating equity from efficiency issues. They hardly mention the idea of designing an abatement strategy based on global costs and benefits and of separating this from the contentious issue of how the burdens might be shared between the participants in an international agreement. Tradeable emission quotas represent perhaps the leading example of an economically efficient, market-oriented approach toward multilateral joint implementation. This option is mentioned only briefly on page 70.

The authors fail to note that existing international policy proposals are far from Pareto-optimal. For example, the Berlin Mandate fails to cut back emissions in the regions *where* it is least expensive, and in the time periods *when* it is least costly to do so. It is a counsel of perfection to insist upon universally acceptable decisions, but thoughtful analysis should at least be able to rule out some bad ones such as those that are prominent in the climate debate.

In this debate, perhaps the most troublesome issues are those relating to equity-efficiency tradeoffs. It is reasonable to propose that the well-to-do OECD nations will have to bear the burden of any abatement that is undertaken during the next few decades. It is quite unreasonable, however, to suggest that all others can be exempted from these requirements over the indefinite future. Sometime during the first half of the next century, it will be essential to obtain participation from the non-OECD countries.

Provided that one is explicit about burden-sharing rules (for example, a gradual transition to population-based rights to emission permits), it ought to be possible to separate equity from efficiency considerations with respect to global climate change. This type of separability could provide a useful background for international negotiations. It sets up the constructive possibility of examining "win-win" situations rather than contentious zero-sum games. These are key ideas, and they are conspicuously absent from Chapter 2.

3

Equity and Discounting in Climate-Change Decisions

Robert C. Lind and Richard E. Schuler

INTRODUCTION

The economic evaluation of climate-change policies is largely based on cost-benefit analysis. The widely accepted standard procedure is to estimate future costs and benefits and to discount them to obtain their net present value. The cost-benefit criterion asserts that a policy with a higher present value of net benefits should be preferred to a policy with a lower net present value. Further, the optimal policy, on the basis of the cost-benefit criterion, is the one that produces the maximum present value of net benefits. Two basic issues have been raised in the context of climate policy and other policy areas, such as energy, that have very long time horizons.

- Is the standard procedure of discounting appropriate?
- If so, what is the appropriate rate of discount to use in the analysis?

These are the two central issues of this paper.

Further, it is a well-known fact that if we discount future costs and benefits using a positive rate of discount, future costs and benefits are given less weight in the cost-benefit evaluation than present cost and benefits. Given a positive but low discount rate, it is true that the present value of a dollar of costs or benefits in the distant future is virtually zero. This means near-term costs and benefits are weighed much more heavily in the cost-benefit evaluation than future costs and benefits, which raises the issue of how discounting and the choice of the discount rate relate to

ROBERT C. LIND is a professor at the Johnson Graduate School of Management at Cornell University. RICHARD E. SCHULER is director of the Cornell Institute for Public Affairs at Cornell University.

our concerns about intertemporal equity. For these reasons Chapter 4 of the IPCC report (1996c) addresses the related issues of intertemporal equity, discounting, and economic efficiency.

It is also a well-documented fact that climate policy will result not only in intertemporal transfers—for example, between people living in the present and near term and people living many generations later—but also in intragenerational transfers among people living at the same time in different parts of the world. Typically, policies to mitigate climate change will cause both intertemporal and intratemporal transfers of resources. Therefore, we need to analyze issues of both intergenerational and intragenerational equity within a common and consistent framework. This paper addresses the material in Chapter 3 on equity and social considerations (IPCC 1996c) as well as Chapter 4 on discounting and intertemporal equity mentioned above.

The most surprising and controversial conclusion of this paper is this: any decision with regard to the mitigation of global warming is more fundamentally a decision about equity and the redistribution of welfare from the present and near-term generations to generations in the distant future and from the developed countries to less developed ones than it is a typical investment decision that can be analyzed entirely using discounted cash flow methods. We argue that there is no discount rate that can be based on either generally accepted principles of economic efficiency or generally accepted principles of equity that, when used in a cost-benefit model, will tell us whether or not we should proceed with mitigation or what the optimal program is. While this is contrary to much of the analysis to date of alternative climate-change programs, it gives further reason to look at the analysis of discounting in the context of equity generally.

Overview of the Paper

Because the scope of this paper is broad and the paper is long, the introduction is designed to give you a road map for what will follow.

First, we deal with the issues pertaining to discounting, the discount rate, and intergenerational equity as covered in Chapter 4 (IPCC 1996c). We attempt to show that the fundamental issue is how to allocate resources across generations and that this issue cannot be solved by optimizing a discounted cash flow equation with any discount rate. Economics, of course, plays an extremely important role in showing us what the trade-offs are, but ultimately, the issue comes down to decisions about the allocation of welfare around the world and across generations over a very long time horizon, and our optimization models, using any discount rate, do not tell us how we should do that. Further, these models seldom deal adequately with the risks of potential future disasters, which the present generation may want to consider in choosing current policies on climate change.

We then address issues that pertain to questions of equity more generally and, specifically, the role of the consideration of equity in both the analysis of public policy and the resolution of public policy issues in the case of climate change. Here we review briefly the material in Chapter 3 (IPCC 1996c), which omits discussion of intergenerational equity and leaves that topic to the discussion of the discount rate in Chapter 4. We assert that in the context of global climate change, these topics of inter- and intragenerational equity must be analyzed together with proper uses of discounting.

The section of this paper on discounting and intergenerational equity begins with a brief review of Chapter 4, Intertemporal Equity, Discounting, and Economic Efficiency (IPCC 1996c). It then proceeds with a discussion of the foundations of cost-benefit analysis and the legitimacy of the compensation principle. The compensation principle is fundamental to one line of argument in support of evaluating public policies and projects using cost-benefit analysis with a market rate of return as the discount rate. We will argue that the characteristics of any problem wherein costs and benefits are distributed over many generations greatly diminish the validity and persuasiveness of arguments based on the compensation principle. This is because there is no way to carry out compensation through cash transfers between widely separated generations; therefore, the rationale for using a discount rate equivalent to the marginal rate of return on capital or the consumers' rate of interest is no longer valid.

Our discussion of discounting and intergenerational equity addresses the justification for the choice of the discount rate based on the conditions for such a rate if the economy is on an optimal growth path. This section analyzes, in some detail, the model put forth in Chapter 4 (IPCC 1996c) and the rationale for both the "descriptive approach" and the "prescriptive approach." The descriptive approach takes the marginal rate of return on capital or the market rate of interest as the revealed social rate of time preference (SRTP) and argues that we should use this market-based rate. The prescriptive approach begins with the conditions for the discount rate on the optimal growth path and divides it into its component parts. This approach then argues that the utility rate of discount should be zero, which implies that the utility of each generation is given equal weight. This approach was first suggested by Ramsey (1928) and has had great appeal for many economists. Following from the assumption that the utility rate of discount should be equal to zero, the prescriptive approach attempts to calculate what the discount rate should be based on the other terms in the equation.

We argue that neither the prescriptive nor the descriptive approach establishes a defensible SRTP for use in cost-benefit analysis. To be defensible, an SRTP should maximize the present value of net benefits to provide the right answer, in any meaningful sense, to the question of what investments we should make in mitigation. Interestingly, although this approach—which employs a utilitarian concept of equity—appears to get around the problem of designing and implementing intergenerational transfers, in the end it founders on exactly that problem. To be policy-relevant, the underlying growth model has to be interpreted as an overlapping-generations model where intergenerational trading through intermediate generations is essential to the relevance of the model and its results.

The trade-off we face with regard to investments in mitigation is between our own consumption today and someone else's consumption many years from now. Again, no market rate or modified market rate indicates at what rate, if any, we individually or collectively as a society would be willing to make that trade-off. The critical factor that drives the analysis to this conclusion is our inability to plan and execute intergenerational transfers over long time horizons.

We then address the serious problem of determining the opportunity cost of displacing higher yield investments with a long-term but lower yield investment, such as the mitigation of potential climate change. This relates to the question of determining the shadow price of capital, but it becomes different if we cannot determine any SRTP that is defensible in our discounted cash-flow equations for consumption. We will discuss some of the factors that determine what the shadow price of capital is and offer a potential solution to this problem using the integrated assessment models that, in most cases, include an economic growth model. We also discuss the need to reformulate our economic decision models to account for risk and uncertainty and take advantage of emerging information.

The third section of the paper, regarding concepts of equity and implications for public policy, addresses equity issues in general as well as the discussion of equity in Chapter 3 (IPCC 1996c). It is interesting to note at the outset the richness of issues and the range of concepts of equity discussed in Chapter 3 (IPCC 1996c), which, while restricted to issues of intragenerational equity, are much broader and more diverse than in Chapter 4, where the utilitarian model was essentially adopted without much discussion of the alternatives as the economist's model of choice.

The section on equity and public policy begins with a brief overview of the IPCC Chapter 3 and describes a wide range of concepts of equity, distinguishing between procedural equity and consequential equity. It attempts to show how each of the different concepts of equity plays out, given the facts surrounding policymaking on climate change. We explain that climate-change mitigation is a public good, that indivisibilities (scale economics and threshold affects) are pervasive, and how this casts even greater doubt on the efficacy of efficiency conclusions, let alone on the efficacy of the welfare conclusions that are derived from cost-benefit analysis using any discount rate.

In the fourth section, Practical Implications for Global Climate-Change Policies, we address how issues of equity should enter the analysis and what role equity may play in negotiating a course of action which will lead to cooperative policies on climate change. Although we believe that equity issues will be central to the policy debate and that we, as analysts, need to clarify the implications of various concepts of equity, the analyst should not make assumptions about which concepts of equity should apply and then present policymakers with prescriptions based on these assumptions unless the underlying ethical assumptions are made clear. Often ethical assumptions are hidden from the decisionmaker or the analysis is presented in such a way that the critical ethical assumptions are not clear to the ultimate decisionmakers. Therefore, the results of the analysis can easily be presented without full disclosure of the ethical premises. This is one of the problems that we have in the case of discounting with a rate based on the "prescriptive approach" where the policy conclusions reached are presented as if they were the equivalent of the analysis of a firm's investment. What happens in some cases is that analysts bury ethical judgments in the analysis and then present the results as generally accepted science.

We also point out, in this section on policy implications, that effective policies with regard to mitigation will require unprecedented international cooperation. Solutions or policies that are close to optimal—as far as minimizing costs of mitigation—may or may not be perceived as equitable. One of the interesting open questions is how important the perceptions of equity will be in the political bargaining that fashions global climate policy. To the extent that issues of equity are important in the context of global climate policy, the typical distinction between issues of equity and issues of efficiency may be blurred: the degree of efficiency that we can achieve will depend critically on the degree to which we can work out arrangements that are both efficient and perceived as being equitable by all the parties. We also discuss briefly the possibility that a system of tradable permits can serve the needs of both achieving procedural and consequential equity, and promoting economic efficiency.

The final section summarizes the significant contributions we believe economists can make in developing analytic methods, policies, and institutions to deal efficiently with global climate problems while making ethical implications transparent.

DISCOUNTING AND INTERGENERATIONAL EQUITY

Chapter 4 of the IPCC report begins with a general and very basic discussion of discounting in the context of climate change and points out that "determining the appropriate discount rate involves issues in normative as well as positive economics" (IPCC 1996c, 130). It then goes on to discuss these issues in terms of a formula derived from optimal growth theory, namely equation (4.1), $d = \rho + \theta g$, "where d is the discount rate, ρ is the rate of pure time preference (also called the utility discount rate), θ is the absolute value of the elasticity of marginal utility..., and g is the growth rate of per capita consumption." It states that "Equation 4.1 provides a way to think about discounting that subsumes many related subtopics, including the treatment of risk, valuing of nonmarket goods, and the treatment of intergenerational equity." The chapter then addresses many of the issues associated with discounting and intergenerational equity in terms of this equation.

In particular, it divides economists into two groups according to how they use and interpret equation (4.1) in arriving at a discount rate for evaluating mitigation options for climate change. One group takes what is called the prescriptive approach, which uses the right side of (4.1) to determine the appropriate discount rate. The prescriptive approach argues that the discount rate, d, should represent the SRTP, or d = SRTP = $\rho + \theta g$, as in equation (4.1), and then proceeds to determine the SRTP by an analysis of ρ, θ, and g. The economists taking this approach generally argue that intergenerational equity requires that ρ, the utility discount rate, be set equal to zero. This would mean that the utility of each generation would be given equal weight in the social welfare function. Then the SRTP would equal the product θg.

Chapter 4 then goes on to discuss estimates of g. It states:

> If the pure rate of time preference (ρ) is zero, then high rates of productivity increase (and thus high g), of the order of 1.5%, plus high (absolute) values of the elasticity of marginal utility (θ) imply a social discount rate of about 3%. With low rates of productivity increase, of the order of 0.5%, and low (absolute) values of the elasticity of marginal utility, the social discount rate is of the order of 0.5%. In a gloomy scenario, in which future output and consumption decline, then g and thus the SRTP may be negative (Munasinghe 1993; IPCC 1996c, 131, 32).

It goes on to state:

> A higher SRTP may apply to developing countries with higher rates of productivity growth. If labor productivity increases by 5

to 8% per year, as experienced by the high-growth countries of Asia, and with an elasticity of marginal utility of 2, discount rates of the order of 10 to 16% could be justified. Similarly, low-income countries close to subsistence levels could have high elasticities of marginal utility (this assumes a rapid fall-off of marginal utility from the extremely high initial levels associated with privation), so that their SRTPs could be high even if they were experiencing slow growth over long periods (IPCC 1996c, 132).

Alternatively, Chapter 4 labels the approach taken by the second group of economists as the "descriptive approach." Basically, this group of economists uses the left side of equation (4.1) to determine the discount rate d which can be extended to read $i = r = d$ where i is the marginal rate of return on capital and r is the consumer's rate of time preference for consumption. They note, "Along a full optimum path, the consumption rate of discount equals the productivity of capital (that is, the social rate of investment; in this case, i_t equals the producer's rate of interest)" (IPCC 1996c, 134). As shown in equation 4A.2, $i = d$, that is, d equals the marginal rate of return on capital (134). Therefore, the economists following this approach use a discount rate that represents or approximates the marginal rate of return on capital. Typically, this rate is taken to be in the 5% to 10% range. This group of economists emphasizes the opportunity cost of displacing higher yield alternative investments when evaluating investments in the mitigation of climate change.

The rest of Chapter 4 is largely devoted to developing the arguments for and against these alternative positions. In addition, Annex 4A, *Methodological Notes on Discounting* (IPCC 1996c, 134–42) provides technical notes and background with regard to these positions.

Chapter 4 is an accurate, concise, and comprehensive summation of the literature, as could be expected given its distinguished authors. The authors, however, did not make a recommendation for either a discounting procedure or the choice of a discount rate, which is not surprising given the state of the professional debate on this issue. Although Chapter 4 is a superb summary of a large and diverse literature, it is important to note several very important areas related to discounting and project evaluation which Chapter 4 does not deal with adequately. Addressing these subjects would have required a much longer paper and perhaps it was felt that these topics were beyond the scope of the assignment.

The first such topic is risk. The authors state:

Most economists believe that considerations of risk can be treated by converting outcomes into "certainty equivalents," amounts that reflect the degree of risk in and investment, and discounting these certainty equivalents (IPCC 1996c, 130).

First of all, it is important to note that in the hundreds of cost-benefit analyses we have come across in our careers, we have never seen one that has systematically converted costs or benefits to certainty equivalents to account for risk. There is a good reason for this. In the case of most public projects or policies it is virtually impossible to do so. Because the costs and benefits in cost-benefit studies of climate-change policy or any other policy are not risk adjusted, it would be inappropriate to use a risk-adjusted or risk-free rate of discount. Therefore, as a practical matter, we have not solved the analytical problem of how to treat risk in cost-benefit analysis even if one accepts the use of certainty equivalents or of procedures used in finance to adjust the discount rate for risk by taking into account how returns from a given investment are correlated with returns in the economy generally.

Further, modern developments in the analysis of investment under uncertainty (Dixit and Pyndyck 1994) and the modern literature in finance have demonstrated that the approach of forecasting costs and benefits, adjusting costs and benefits for risk to obtain certainty equivalents, and then discounting at a risk-free rate is not the optimal procedure for the analysis of investment decisions under uncertainty. What is required is a sequential decision process which takes advantage of obtaining information over time. In this sequential decision process, one does not make a once-and-for-all, go or no-go decision, but rather decides what the next step should be in order to respond to the situation at the next decision point. This process resembles buying an option. Discounting is still involved, but the entire framing of the decision problem is changed. We believe that this approach is particularly well suited for the analysis of how we should respond to potential climate change. A few cost-benefit studies have used a dynamic decision-analysis framework to calculate the value of information. However, they are not typical. Further, almost all cost-benefit analyses do not adjust the discount rate for risk or account for risk aversions or risk preference, nor do they address what, in our minds, is the most significant risk associated with climate change, namely the risk of truly catastrophic outcomes.

The authors fail to address one other topic in any detail: how in practice one measures the opportunity cost of displacing investments that have high rates of return but relatively short life spans with investments that have much lower rates of return but produce this return over much longer time horizons. The authors of Chapter 4 are clearly aware of the issue of displacement. They state:

Nordhaus (1994), Lind (1994), Birdsall and Steer (1993), Lyon (1994), and Manne (1994), among others, have all stressed the importance of the opportunity cost of capital, noting that even apparently small differences in rate of return result in large differences in long-run results. Over 100 years, an investment at 5% returns 18 times more than one at 2%. Thus, where some redistribution of future returns is possible, society would be foolish to forgo a 5% return for a 2% return.

Birdsall and Steer of the World Bank (1993) explain the need to direct investment to the most productive uses, warning against use of too low a discount rate (IPCC 1996c, 132).

However, the problem is that economists who stress the importance of not displacing high-yield investments with a lower yield one frequently do not address how to analyze the opportunity cost involved, or they imply that using a rate of discount equal to the marginal rate of return on capital will correctly solve the problem—which is only true under very restrictive conditions where the economy is fully optimal.

Consider a very long-term investment, say in mitigation, which earns a 2% rate of return. This investment is equivalent to investing at 2% today and continually reinvesting the principal and interest until the investment pays out centuries later. Higher yield standard investments in the economy pay out over a much shorter time period, and the returns go back into the economy. On average, most returns are consumed rather than reinvested. The technique for handling this problem, as Chapter 4 explains, is to convert all spending, including investment, into consumption equivalents (IPCC 1996c, 130). The technique for doing this in the case of investment is to multiply investment by the shadow price of capital (131).

As a practical matter, this is seldom done because of the difficulty of determining which costs will displace investment and which will displace consumption, and the even greater difficulty of predicting which future benefits will take the form of investment or consumption (Arrow 1996; Lind and others 1982, Chapter 2). Further, there is little agreement on the value of the shadow price of capital, with estimates that can range from near one to infinity depending on assumptions about the marginal rate of return on capital, the discount rate applied to consumption, the length of the investment alternatives displaced, and the fraction of investment returns that are reinvested. If there is no defensible SRTP, as we will argue, then the problem becomes even more difficult because one cannot—even in theory—compute a shadow price of capital (IPCC 1996c, 131). These issues will be addressed briefly in this review.

The Cost-Benefit Criterion and Discounting in an Intergenerational Context

Cost-benefit analysis was designed to extend the criteria of profitability, which governed the allocation of resources in the private sector and produced Pareto-optimal outcomes in perfectly competitive markets to cover the case of projects in the public sector where there was either a public good or significant externalities that were not taken into account by the market. Cost-benefit analysis in its purest form assumed that the starting point was an economy in competitive equilibrium and that the economic conditions in that economy were consistent with perfect competition. The basic idea was to measure all costs and benefits in terms of willingness to pay or willingness to accept payment, then to calculate the total net benefits and discount them to their present value using a rate of discount equal to the market rate of interest. In a perfectly competitive world, without distortionary taxes, there would be a single risk-free rate of interest equal to the marginal risk-adjusted rate of return on capital. Because everyone could borrow or lend at this rate, this rate would also equal the individual's marginal rate of time preference for consumption according to traditional neo-classical economic theory. This is well understood and does not need to be elaborated further.

If the cost-benefit criterion were met and the present value of net benefits were positive, it followed that the people who received the benefits could compensate the people who had incurred the costs, a surplus would remain, and adopting a project could result in a Parto improvement. This is known as the Kaldor-Hicks hypothetical compensation test. If this test is met, one can, in principle, create a Pareto improvement where some or all people are better off and no one is worse off. In the case of a public project, satisfying the Kaldor-Hicks compensation test does not necessarily mean that the project would bring about such a Pareto improvement unless compensation is actually paid. In the case of private projects, full compensation is paid because the costs have to be covered by revenue. We can therefore say that if a public project passes the cost-benefit test and compensation is paid, then there will be a Pareto improvement. If full compensation will be paid, satisfying the cost-benefit criterion is a sufficient condition for undertaking a project.

On the other hand, even if compensation will not be paid, a strong case can be made for accepting only those public projects that meet the cost-benefit criterion. The argument for adopting the cost-benefit criterion, even for cases where compensation might not be paid, is this: if one wanted to subsidize the beneficiaries of a project, one should only undertake the project if the net benefits are positive; otherwise the beneficiaries could be made equally well off as they would have been with the project by giving them a cash payment that is less than the cost of the project. Notice that the decision to undertake such a project has two decision components. First is an ethical decision (made politically) as to whether we want to make the transfers that the project without the payment of full compensation entails. Cost-benefit analysis tells us nothing about how to make this decision. The second decision is how best to carry out this transfer. Should we put in place the project or should we make cash transfers? If the project satisfies the cost-benefit criterion, then it is more efficient to make the income transfer by undertaking the project; otherwise we should make cash transfers. In this case, satisfying the cost-benefit criterion is a necessary condition for undertaking the project.

Suppose we are considering a project or policy which does not meet the cost-benefit criterion and where, for whatever reason (political or otherwise), cash transfers cannot be used. Would we ever want to undertake the project or policy anyway? In other words would we ever want to make a transfer using a project which is less cost-effective than a cash transfer would be, but where the alternative of making the cash transfer is not an available option? The answer is clearly yes, but it hinges on the fact that we want to make a particular transfer—that is a value judgment independent of the cost-benefit criterion.

The fact that cash transfers cannot be used means that some projects or policies—although more expensive than appropriately designed cash transfers—will be used to effect these transfers. This happens all the time in situations where political forces dictate that we want to undertake a project, such as a water project or a trade policy where the costs exceed the benefits, but where cash transfers are an unacceptable or infeasible mechanism for obtaining the objective. One might argue that this happens as a result of some group's political power and has nothing to do with ethical considerations. We take the position that the political system and political bargaining determine, de facto, society's values with regard to ethical decisions on income distribution. Therefore, these are the ultimate ethical decisions or valuations.

In addition, there are several other arguments for following a policy that approves only projects that satisfy the cost-benefit criterion with or without compensation. First, some would contend that with a large number of projects, on average, everyone would benefit if we undertake only projects that satisfy the cost-benefit criterion. In other words, while any one project might not produce a Pareto improvement, all projects taken together will. However, a look at the history of public works projects and other projects that were subjected to cost-benefit analysis suggests that certain sectors of the country and groups of people benefited disproportionately from these projects (for example, the water projects in the western part of the United States). Even if the benefits and costs do not average out on an individual basis, the government can use other redistribution mechanisms to address resulting inequities. A strong argument can be made that the best policy is to pursue only those investments that meet the cost-benefit criterion using a market-based rate of discount and to deal with any inequities that come about through tax policy or other redistribution mechanisms.

The foregoing logic runs into some difficulty when we look at the cost-benefit criterion in terms of investments that have very long time horizons and where the costs are paid by generations living today and in the near future and the benefits go to the generations living far in the future or conversely. In the case of such an investment, if we decided that we could more cheaply compensate people in the distant future for the loss of utility as a result of climate change by giving them cash, we would somehow have to set up a trust fund today and invest and reinvest that money until, at some future date when global warming was upon us, we could give the future generation additional consumption goods that would compensate them for the lost environmental amenity, if that is possible. The problem, which has been discussed by Cline (1993), Lind (1995), and others, is that we don't know how to go about setting up such a trust fund. In the case of global warming, we would somehow be required to set aside special resources on a worldwide basis and continue to invest and reinvest them at the going market rate of return until the time that compensation was to be paid. In addition to that problem, there are the problems of making sure on an international basis that these resources would go to the people in the future who would actually incur the damages, and of making sure that the value of these resources was not diminished by global warming itself (Lind 1990). For example, investing in Bangladesh to provide future generations with resources in compensation for the effects of climate change would be futile if Bangladesh were under water as a result of global warming.

Furthermore, another problem greatly increases the problem of compensation: namely, that such a program would require the cooperation and commitment of intervening generations over the period until compensation was to be paid. There is no way to commit future generations to a continuation of the program, and we might not want to—even if we could—because that would restrict their ability to adapt policies to new information and changing technology. Some future generation would have every incentive to simply appropriate the capital that had been set aside and turn it into their own consumption. Therefore, even if the current generation were committed to setting aside resources to be invested for the benefit of future generations who may incur the cost of global climate change, there would be no way in practice that this could be carried out. It should be noted that the precommitment problem is not just a problem associated with intergenerational income transfers.

Any long-run program requires the commitment of intervening generations, including mitigation itself; although the temptation to consume a large store of resources that have been set aside may exceed the temptation to discontinue a program. This has sometimes been referred to as a time inconsistency problem. We might ask why, if we are willing to make such a commitment, we shouldn't expect future generations to do the same. A full discussion of this is beyond the scope of this paper, but a few reasons are that they may have different values, face different circumstances, have different information, or have greater incentives for breaking the commitment.

Similarly, if a project met the cost-benefit criterion, a series of transfers from future generations to present ones could, in theory, be set up through intervening, overlapping generations. Such transfers could ensure that future generations—who are the beneficiaries of this investment in mitigation—could transfer resources through intervening generations to the current and near future generations, who would pay for mitigation abatement; both present and future generations would be better off than they would otherwise have been. This would be a case where full compensation would be paid and there would be a Pareto improvement across all generations. Lind (1995) discusses this process and how it might work in some detail. In a world where we would expect future generations to be better off than the current one, this scheme for paying compensation would be both efficient and equitable in that the richer generations of the future would compensate the poorer near-term generations for the benefits they received from the current investment. The basic point is that, in theory, one could use intergenerational transfers—carried out through overlapping, intervening generations—to carry out compensation schemes in conjunction with choosing policies that met the cost-benefit criterion, and the result would be both efficient and equitable. As stated by Lind (1995):

> ...before adopting the cost-benefit criterion for evaluating global warming policy, we have to investigate the feasibility of implementing such an intergenerational compensation scheme as a practical matter. In many ways, the most interesting intergenerational issue is whether we can structure intergenerational transfers or trades so that we can then pursue efficient investment strategies to maximize the consumption stream across all generations and then take care of any inequities through intergenerational transfers.

The rationale for using a discount rate equal to the marginal rate of return on capital in conjunction with the cost-benefit criterion is this: if one were to be able to invest and reinvest at a rate of return equal to the return on capital, then one could always create a payout scheme through an investment that would dominate, even for a very long period of time, any investment that had an internal rate of return that was lower than the marginal rate of return on capital. Similarly, if it turned out that, because of market imperfections such as those caused by the income tax, the marginal rate of discount that individuals applied to their own consumption in the future was below the market rate of interest, then it would also be true that, if one could undertake transfers from future generations back to preceding ones to compensate them for the costs they incurred for mitigation, it would be possible to compensate everyone if the investment in mitigation had a rate of return above the marginal rate of time preference of the individuals who over time incurred the costs and engaged in the transfer process.

The problem is that, in the case of a public investment taking place over many, many generations, it is impossible to design the appropriate set of transfers in either direction and commit to their being carried out. This does not mean that intergenerational transfers cannot occur. They can and do occur all the time. What it does mean, though, is that one cannot make investments with any possibility of carrying out a planned reallocation of the returns across generations, over time, to deal with problems of equity. The logic of cost-benefit analyses and the appeal of the cost-benefit criterion requires that we be able to make trades across generations at known rates of return or discount that make it defensible to convert benefits and costs over time into a single number which is their present value.

Unfortunately, the answer to the question posed by Lind (1995) is that we cannot plan and execute with confidence such transfers that would justify adopting the cost-benefit criterion using a market-related rate of discount. Also, it is important to note that the marginal rate of return on capital and the consumers' rate of time preference for their own consumption are relevant to a trade-off between consumption by those living today and consumption by people living many generations in the future *only* if such transfers between non-overlapping generations can be carried out through intervening generations.

In the case of climate change, all we know is that, if the present and near future generations invest a certain amount, they can give benefits to generations in the distant future. Whether we consider this a wise idea or whether we would want to go ahead with such an investment depends purely on whether we would want to make this transfer. This is not an investment in which the present generation can calculate whether its wealth goes up as a result of making that investment. Rather, it is a pure gift from the present generation to distant future generations. Therefore, it is much more like a charitable gift or foreign aid (Schelling 1995). Obviously, whether one wants to make such a transfer depends on economic factors, such as how much it will cost, how much the recipients will benefit, and how the money will be used. But the decision has little to do with whether the investment passes the cost-benefit test given any rate of discount, whether positive, zero, or negative.

Under these circumstances, there is no reason to believe that the value we put on the benefits or costs to far future generations should be discounted according to some exponential formula; and it is perfectly reasonable to expect that every individual looking at this question may have a different rate of discount. This is a pure redistribution problem, and cost-benefit analysis—using any rate of discount—does not tell us whether we would want to make this transfer. It would be exactly analogous to using cost-benefit analysis to ask, do we want to give foreign aid to Brazil for investment in education? It may be that if we calculated the benefits, costs, and the internal rate of return, the return would be extremely high, say 30% or 40%—which is far better than most investments that we can make in the United States. Yet, we might decide that this is not something we want to do. We simply might not want to transfer to Brazil the resources needed to educate Brazilians. The reasoning has nothing to do with the rate of return on the investment or on market rates in the United States.

Obviously, the rate of return on the investment is relevant and may influence our decision in the sense that, if we could invest a dollar and give value to the Brazilians that was worth billions, we might choose to make that transfer. But we would not apply any standard kind of investment calculation. For this reason we would probably want to display the internal rate of return on investment in education in Brazil, but we would not base our decision on the present-value calculation alone.

This example is worth exploring a little further. Clearly, if the benefits from an investment in Brazilian education could be captured, assuming a 30% rate of return and compensation paid to American investors whose opportunity cost of capital was 5%, then the investment should be evaluated like any other investment using a 5% discount rate, that is, the U.S. rate. Suppose, however, the benefits, for whatever reason, could not be captured so this investment was a pure gift to Brazil which we decided to make. We still might ask ourselves, are we making the best investment for the citizens of Brazil? Suppose the marginal rate of return on capital in the private sector in Brazil was 20%. Then we would want to do a cost-benefit analysis using a 20% rate of discount to determine whether, perhaps, we should just give the Brazilians the money to invest in their economy. Further, suppose that instead of investing in Brazilian education at 30%, we could invest in public health in Brazil and get a rate of return of 50%. Clearly, this rate would be the appropriate measure of the opportunity cost of any investment for the benefit of Brazilians generally.

One lesson emerges from this simple example for evaluating climate-change alternatives. If you are investing in the United States, the benefits are going to Brazil, and you have made an ethical decision to undertake this transfer, then you should also take into account the fact that you could give them the cash value of the project instead of making that investment. This means you should consider other investments in Brazil that would also benefit Brazilians, including investments in the public sector such as public health. This discussion reflects one of the criticisms most frequently heard with regard to climate-change programs that will largely benefit the tropical regions of the third world: the money could instead be invested in other public or private projects that would provide many times the expected rate of return from climate-change mitigation.

Optimal Growth and Discounting:
A Prescriptive and Descriptive Interpretation

The second approach to the determination of the discount rate follows from the model of optimal economic growth. Economists frequently appeal to this approach in their search for a discount rate to use in models designed to evaluate alternative mitigation policies, and it features prominently in Chapter 4 (IPCC 1996c). The basic line of argument is as follows. If one takes the model of optimal economic growth, it follows that on the optimal growth path, i, the marginal rate of return on capital equals r, the consumer's rate of discount with respect to consumption, and these two equal the SRTP, the social rate of time preference. It also follows that the SRTP equals the sum of ρ and θg where ρ is the utility discount rate, θ the absolute value of the elasticity of marginal utility, and g is the growth rate in per capita income.

$$i = r = \text{SRTP} = \rho + \theta g \tag{1}$$

It is then assumed that the economy is at a point on the optimal growth path so that these conditions hold. Of all the variables in equation (1), we can empirically estimate i, the rate of return on capital, and g, the growth rate of per capita income. We can speculate about θ, the absolute value of the elasticity of marginal utility. However, we cannot empirically verify the value of ρ.

As previously discussed, two alternative approaches have been proposed for determining the discount rate using equation (1). Consider first the prescriptive approach. It begins by looking at the right-hand side of equation (1) and observing that, in the optimal growth framework, inter-generational equity requires that ρ, the utility rate of discount, should be equal to 0. If so, every generation's utility would be given equal weight.

In that case the SRTP, or the discount rate for discounting alternative consumption paths, would be equal to θg. Given assumed rates of g in the range 0.5% to 1.5% and standard estimates of θ between 1 and 2, the SRTP falls in the range 0.5% to 3.0% (IPCC 1996c, 131–32). It is important to note that any reasonable estimate of the marginal rate of return on capital is significantly above this range of estimates of the prescriptive rate.

This implies that if ρ is 0 and our range of estimates of θ and g are correct, then we are not on the optimal path because the implied rate of time preference is significantly below the marginal rate of return on investment. In addition, it implies that the current generation saves and invests far too little. As Chapter 4 (IPCC 1996c, 133) points out, work by Manne (1995) has shown that with a marginal rate of return on capital of around 5% and an SRTP in the 2% to 3% range, an enormous increase in near-term savings and investment would be required to bring the economy back onto the optimal path. In fact, in the short term, the required increase in savings and investment might be much higher than any increase that we could imagine taking place or being tolerated by the current generation. Furthermore, an SRTP of 3% would imply that undertaking investments with a low 3% rate of return, including climate-change mitigation, would not necessarily be the appropriate response. Before doing that, we should undertake instead all the investments with higher rates of return until we got down to investments with 3% rates of return. The available pool of savings might well have run out before that point had been reached. The IPCC report (1996c) suggests that we do not have more investment because of market imperfections. However, we don't believe that eliminating corporate and personal income taxes would lead to the massive increase in savings that would be required to justify investments with a 3% rate of return.

The alternative descriptive approach takes the position that the variable i, the marginal rate of return on capital, which we can observe, should be used as the SRTP in our discounted cash flow models of costs and benefits associated with mitigation investments.

Without going into more detail with regard to these two approaches, it seems to us that several fatal flaws in both approaches invalidate the use of a discount rate obtained by either. First of all, both approaches implicitly assume that the United States and the rest of the world accept the utilitarian formulation of equity and would be willing to apply it in making decisions with regard to intergenerational transfers. In the intergenerational context, and in the presence of per capita income growth, this assumes that the poorer present and near-term generations should be transferring resources to the richer future generations. This concept, in itself, should make us pause. Note another ethical system that has drawn the widespread attention of economists, that of Rawls (1971), which would imply that richer future generations should be making transfers to the poorest generation—that is, the present one.

The utilitarian framework can also be applied to intragenerational transfers and utilitarian intragenerational equity. This would imply that we should be making transfers from rich nations to poor nations and rich people to poor people until we have equalized the distribution of income. Many, if not most, economists would subscribe to the view that intergenerational equity is probably good and that a reasonable definition of intergenerational equity within the context of an economic growth model would mean setting ρ equal to 0. But if these same economists were faced with a vote on whether we ought to go ahead with an equivalent program of intragenerational equity and transfer resources from the rich to the poor until equalization occurred, very few indeed would vote for such a program if they thought it had any chance of passing. Clearly, individuals and society as a whole—in making intratemporal, or for that matter intertemporal, decisions—do not behave as if they were pursuing the utilitarian principle of equity. One of the advantages of considering intra- and intergenerational equity within the same context is that it points up gross inconsistencies that economists, of all people, should wish to avoid. At the very least we can conclude there is insufficient support for the utilitarian model of distributive justice among decisionmakers to justify making it the ethical pillar of public policy.

The prescriptive approach might be defensible if the economist, instead of assuming ρ = 0, would explain very carefully the underlying assumption and many of the implications of such an assumption as part of their analysis. Obviously, as individuals, we do not give equal weight to changes in other people's utility compared with our own. Nor do governments give equal weight to the utility of people in other countries as they do to the people at home. If they did, rich countries would engage in massive foreign aid programs, and they do not. As Schelling (1995) points out, people do not behave as if they give equal weight to the utility of themselves, their children, and their distant descendants. Given that programs to mitigate global warming, in all probability, will transfer utility from near-term generations living in industrialized countries to distant future generations, it is wildly implausible to believe that the public and policymakers in the United States today would weigh the costs and benefits realized by U.S. citizens today equally with the costs and benefits of people living in India two hundred years from now.

Therefore, aside from other conceptual problems, the prescriptive approach to discounting and the determination of the discount rate is based on ethical premises that are totally inconsistent with those of the public at large and government decisionmakers. To carry out an analysis of climate policy using cost-benefit analysis and the prescriptionist approach to discounting and then to present it as standard investment analysis could be viewed as downright sneaky. In summary, the ethical premises are clearly wrong, and they are buried in the analysis.

A second reason shows both approaches to be badly flawed: there is limited evidence that economies are on an optimal economic growth path. Therefore, the conditions in almost any economy probably do not correspond to the conditions set forth in equation (1). This is particularly damaging to the prescriptive approach, which relies so heavily on the relation between the SRTP and ρ, θ, and g in the right side of equation (1).

The fatal flaw with the descriptive approach is that it is predicated on the assumption that intergenerational trades at market rates can be made both ways. Chapter 4 states that

> The descriptive approach assumes compensation from one generation to another for any loss of environmental amenities, implicitly leaving unanswered whether compensation is likely to occur. The descriptionist view argues for choosing the path that maximizes consumption, making transfers separately out of the larger present value of consumption (IPCC 1996c, 133).

Another argument in support of the descriptive approach states that "the appropriate social welfare function to use for intertemporal choices is revealed by societies' actual choices; advocates of the descriptive approach generally call for inferring the social discount rate from current rates of return on growth rates" (IPCC 1996c, 132). While we are sympathetic with these sentiments, the discount rates in the optimal growth model are instantaneous rates that apply to people's trade-off between their own consumption now and their consumption at an infinitesimally close time from now. There is no way this can be interpreted as the rate at which someone living today (assuming an overlapping-generations model) would be willing to trade his or her own consumption now for the consumption of someone else two hundred years from now.

Unfortunately, the conclusion we must reach is that we have no valid way, based on observed market data, to tell what the trade-off would be or should be—from the standpoint of any individual—between his or her own consumption now and the consumption of someone else living in the distant future. The same is also true for societies as a whole. In other words, neither market data nor economic principles of deriving a defensible SRTP can be applied to choices between consumption by generations separated by long time intervals.

Schelling (1995) reached the same conclusion by looking directly at the trade-offs involved. However, Schelling's conclusions do not hold if the appropriate systems of intergenerational transfers can be put in place.

In that case the rate of return on investment over time does matter, as do individual rates of time preference for one's own consumption at different points in time. And these trades in time between generations can be analyzed as standard investments using discounted cash flow models. In this case, one's preferences for one's own consumption compared with someone else's consumption at a different place or time only matters in the cost-benefit framework used to decide how we distribute the net benefits and whether we require that compensation be paid.

What Is the Role of Economic Analysis in Determining Climate-Change Policy?

The question immediately arises. If we cannot do standard cost-benefit analysis and use the cost-benefit criterion based on discounted cash flows because there is no rate of discount that we can defend for use in the standard cost-benefit model, what then should economists do? Clearly, the decisions about when and if to undertake mitigation investments to prevent future climate change will depend heavily on what the costs and the benefits are over time. So the activities of economists and others— namely, calculating costs and benefits over time—remain as useful as ever. But because this problem takes place over many generations in an environment where it is not always possible to make intergenerational transfers, it is very important that we describe and display the cost and benefit data over time.

We do this now for climatic effects, but most cost and benefit data are presented as one number, present values. Therefore, instead of presenting a present value for costs and a present value for benefits or a present value of consumption, the appropriate display of these estimates would be graphs over time so that we could see how the various significant economic variables change over time. After all, the problem is one of intertemporal reallocation. For decisionmakers to make informed judgments, it is extremely important that they be able to see the time profile of various economic variables, not to be presented with simply a present value. Therefore, all of the activity by economists associated with estimating costs and benefits over time is every bit as valuable as if it were going to be the input to a maximization model designed to maximize the discounted present value of net benefits.

Quite frankly, we have been frustrated in looking at the output of integrated assessment models, where often the data is presented in terms of present values rather than showing the time profile of costs and benefits over time—even though the models generate this data. For example, one problem with this approach is that with the assumptions built into the economic models, the future per capita consumption of generations living several hundred years from now—when the impact of global change is predicted to become significant— is around twenty to thirty times the per capita consumption today. This is true assuming even rather modest rates of growth in per capita consumption in the range of 1% to 1.5%, which is standard in these models. Anyone who understands the economics and the underlying assumptions understands this.

However, to clarify the concept for a more general audience, graphs showing this consumption growth over time would make it absolutely clear that future generations—according to our economic assumptions— are going to be much, much wealthier than we are today. This increase in wealth more than compensates for even the worst estimates of the loss of economic value resulting from climate change. We may or may not agree with this conclusion, but at least everyone would be aware of the implications of the underlying assumptions about future economic growth. This is buried in the analysis when we display costs, benefits, and net benefits as present values and when we present only what we have calculated to be the optimal program.

In summary then, our first recommendation to the economists and integrated modelers is this: in analyzing various economic and other physical dimensions of the climate-change issue, the results should be displayed as time profiles in addition to present values. This will lead to improved information for decisionmaking because the distribution of these variables over time is critical to the decision.

It is also critical to decisions regarding whether to invest in mitigation to determine what would be the stream of benefits if these same resources were invested in normal capital markets instead of in mitigation. In our view, one of the biggest challenges to economic modelers is how to capture and display the opportunity costs of foregone alternatives. What would be the alternative benefit streams in terms of consumption if we made comparable investments in public health and education in the Third World, which might yield many times the rate of return on mitigation. One way to calculate the opportunity cost of displacing marginal private investments, using the growth models that are part of the integrated assessment models, would be to assume that all of the resources aimed at mitigation could be redirected to investment in the private sector at private-sector rates. Suppose we believe that of the resources required for investment in mitigation, 20% would be drawn from private investments and 80% from consumption. If all these resources were channeled back into investment, then there would be an increase in total investment in the economy over what there would have been without any mitigation program at all, equal to 80% of the investment budget for mitigation. Put differently, the 20% that would be taken out of private investment would be left in private investment and the

80% taken from consumption would be put into the private capital markets and invested at competitive rates. Note that one set of runs that one would want to make would be to assume these resources were invested in the Third World at the high rates that can be obtained in these countries. This would mean using a growth model for the Third World and tracking the alternative consumption streams.

If we performed this experiment, we could determine what the rate of economic growth would be if the resources that were earmarked for mitigation were put into regular investment channels. This still would not capture the opportunity cost of not investing in high-yield public investments such as health and education, but it would be a step in the right direction. Note, this higher economic growth and higher consumption over the long run would accrue not only to generations in the far future when global warming would occur, but also to all of the intervening generations. The difference between the two consumption streams—one with mitigation investment and the other with the resources, which would have gone into mitigation, invested in the private sector—would give us a measure of the opportunity costs over time associated with investing in mitigation rather than in general economic development.

Any economist doing this work will obviously feel a strong urge to discount the difference in the consumption streams to a present value. But given the previous discussion, the corresponding value will be essentially meaningless. This doesn't mean that we may not want to compute some present values on occasion, but we must understand that they are simply illustrative. And they may be helpful to decisionmakers who, like most economists, like to think in terms of present values.

Our economic models still have a great deal of value, but that value is in simulating alternative scenarios and not in optimizing, because we have no way of determining the intertemporal weights that would allow meaningful optimization to take place. As economists, while we cannot tell decisionmakers what the optimal decision is, we can give them a great deal of information about the behavior of variables, which they need to know to make any decision. In fact, this is where the value of cost-benefit analysis really lies: in informing decisionmakers of the various trade-offs involved so that they can make an intelligent decision.

We make these recommendations with some trepidation as we realize that optimization has become synonymous with economics itself and with economic modeling. Unfortunately, in many cases we don't have the information required to determine an optimal policy. We would do better to focus our attention on providing information that is informative to decisionmakers rather than trying to create a complete criterion function by which we attempt to tell them what the decision should be.

Uncertainty and Sequential Decisionmaking

As we pointed out, one area of Chapter 4 (IPCC 1996c) that we felt does not reflect the most current thinking on investment decisions under uncertainty was the treatment of uncertainty and risk. Using the old approach to cost-benefit analysis, or investment analysis for that matter, we would try to estimate the expected benefits and expected costs, adjust them for risk to a certainty equivalent along the lines suggested in Chapter 4 (130), look at the net present value of benefits using a risk-free discount rate, and make a go or no-go decision on the investment depending on whether this number was positive. In other words, we would consider all the factors relevant at a given point in time, adjust for factors relating to risk, and then make a decision. The modern literature on investment decisionmaking under uncertainty and on modern finance recognizes that information becomes available through time or can be produced; therefore, the way to approach these decisions is to make them sequentially, realizing that at each point in the process one is essentially buying an option to proceed with the next step of the process. This approach is probably best explained in Dixit and Pyndyck (1994).

This sequential approach to decisionmaking under uncertainty is probably the most important point to get across to decisionmakers working on the problem of climate change. We should not be sitting back and trying to do a definitive cost-benefit analysis over a four-hundred-year period to determine whether or not we should embark on a mitigation project today. What we should be doing is looking at the evidence, looking at what information we think we might be getting over the next twenty to thirty years, and setting a planning horizon that is tractable, anywhere from ten to thirty years, over which we will undertake a course of action that will put us in a position to move ahead with further mitigation investments in the future if we choose to do so.

Perhaps it is fortuitous that it is impossible to do definitive cost-benefit analysis over long time horizons where intergenerational transfers cannot be designed and implemented. We probably should not have been trying to do them in the first place, although it is very important to try to estimate the long-run potential for climate change and the long-run costs associated with that potential. If in fact we break up our policy horizon into shorter time segments, then we also solve many problems of the discount rate. For example, we can talk with some confidence about the costs of abatement over the next thirty years, and within this time frame it is perfectly appropriate to discount costs and benefits using a market-related rate of discount. So, while it may not be appropriate to use net present value calculations in summarizing streams of costs and benefits over a very long time horizon, this could be unimportant. If we frame the policy decisions appropriately, we never have to discount over very long periods. Rather, we will be making decisions sequentially about what we want to do over the next time horizon and within these time horizons we can use standard discounted cash flow calculations to summarize costs and benefits. For the purposes of these decisions, the relevant benefits will be how much the current generation is willing to pay to give future generations better options for coping with the potential effects of climate change.

CONCEPTS OF EQUITY AND IMPLICATIONS FOR PUBLIC POLICY

Chapter 3 (IPCC 1996c) discusses comprehensively intragenerational equity issues as they pertain to global climate change (GCC) and to policies to prevent or offset the effects of its occurrence. Yet this discussion (summarized briefly in our section, Perspectives on Equity, below) offers little guidance on how, if we reject the discount rate apparatus for comparing impacts that span generations, formal concepts of equity might be substituted and used to establish coherent sustainable long-range policies. The equity problems we encounter are simultaneously intergenerational and international. Not only do we worry about the effect of our actions on our own nation's future generations, we are also concerned about the consequences of the current actions of other nations on our future generations and of their future generations' actions on our future generations.

As illustrated in Figure 1, traditional discount rate calculations attempt to weigh and compare impacts like E_{12}^A and E_{12}^B that flow over time. Even within a single country or trading group, if the time horizon spans generations, we have questioned the desirability of using this method. Its use is even more problematic if the impacts are simultaneous across cultures and generations like E_{12}^{AB} and E_{12}^{BA} (my country's contributions to global climate change adversely affect some other nation's future generations). And trying to estimate now how future generations in different countries might affect each other, let alone making a welfare assessment of those estimated impacts (E_2^{AB} and E_2^{BA}) is impossible. The primary inferences we might make from historic and current behavior relate to attitudes toward contemporaneous cross-cultural impacts (E_1^{AB} and E_1^{BA}). In these cases, the evidence of nations willingly transferring wealth to effect international impacts is usually quite small (international aid), except in instances like World War II and the Cold War when a nation's very integrity was perceived to be at risk.

By comparison, the case of possible GCC involves small current threats to national survival; for most nations the threats are uncertain and

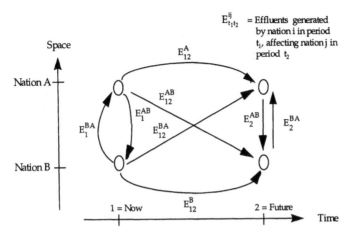

Figure 1. Potential Flows of Adverse International and Intergenerational Environmental Effects.

far in the future. Thus, the primary emphasis needs to be placed on deducing values for intergenerational equity and on making international equity comparisons far in the uncertain future. The primary case for considering contemporaneous equity with respect to GCC is that it may become a prerequisite for establishing an adequate apparatus for systematically dealing with those problems in the future.

This section begins with a brief overview of the relevant equity perspectives. We then discuss how these problems of equity allocations and their consistency are complicated when applied to problems of GCC and sustainability, using the discount apparatus. The ethical difficulties are compounded further when we deal with a public good, like reducing the prospects of GCC, and when significant indivisibilities (threshold effects and scale economies) are associated with consumption and production, as in the cases of GCC and of many proposed impact-reducing actions.

Perspectives on Equity

The primary conceptual equity distinctions made in Chapter 3 of the IPCC report are between *procedural* equity, which focuses on how the views and welfare of affected parties influence the deliberation process and its dynamics, including whose voices are heard and in what proportions, and *consequential* equity, which compares and weighs differential outcomes (IPCC 1996c, 85-87). Consequential equity has been gauged from a variety of perspectives: parity, proportionality, priority, utilitarianism, and Rawlsian distributive justice. *Parity* and *egalitarianism* are closely related, requiring an equal distribution of burdens or benefits. They differ from *Rawlsian distributive justice,* which strives for an equal end result and therefore can allow an unequal distribution of benefits as long as the disadvantaged members of society move toward the level attained by others. *Priority* assigns the benefits to those with the greatest need and so, in a dynamic sense, can move toward Rawlsian distributive justice. By comparison, *utilitarianism* distributes the burdens and benefits in ways that provide the greatest good for the greatest numbers without regard to the extent of inequality (this is the welfare criterion implicit in any discounted net present value sum of benefits and costs over time). Note that the authors of Chapter 3 do not rush to embrace utilitarianism as the equity concept of choice upon which to base GCC public policy—as many economists do when considering intergenerational equity. Finally *proportionality* distributes burdens and benefits in relation to the contributions of claimants.

The remainder of the equity discussion in the IPCC Chapter 3 is divided into a review of pertinent international laws governing environmental degradation with an eye toward their equity implications, a summary of historic cumulative contributions to potential GCC by various nations, a discussion of likely future contributions by various nations, and finally a discussion of the policy implications of the above and their equity consequences. Chapter 3 emphasizes the importance of equity considerations, both ethically and pragmatically, as a means of gaining widespread participation. In particular, it emphasizes the role of industrialized nations, as the primary historic source of current accumulations of GCC potential, and the need to allow the contributions to GCC of lessdeveloped nations to grow as their economies expand. Nowhere is there any mention of the potential future GCC burdens some rapidly developing countries might place on other nations because of their enormous current and projected population levels.

The report acknowledges the importance of relating requirements for compensation to the knowing infliction of harm and suggests that it is important that the injured nation be willing to accept the compensation. Nevertheless, the report also stresses that the implications may vary systematically between developed and developing nations, with an emphasis on the less-developed regions. This special attention to poorer nations and their problems is based on arguments of both consequential equity and procedural equity, since institutional strengthening may be required to help those nations participate fairly. Chapter 3 also advocates transparency of process, particularly as a way to assist poorer nations gain understanding. Finally, the authors express concern about the impact of policies on the distributional patterns within individual nations.

An important conclusion of Chapter 3 is found in the following statement:

Scientific analyses cannot prescribe how social issues should be taken into consideration and how equity should be applied in implementing the convention, but analysis can clarify the implications of alternative choices and their ethical basis (IPCC 1996c, 118).

The report also states: "Equity among generations (or intergenerational equity) is the subject of the next Chapter [4]" (IPCC 1996c, 85). Yet Chapter 3, by emphasizing pursuit of the goal of sustainability, as laid out in the Framework Convention on Climate Change (FCCC), does advocate one measure of intergenerational equity.

Implications for Discounting and Intergenerational Equity

Although consequential equity measures compare alternative states, they can be used for both dynamic and contemporaneous comparisons. As an example, if two current societies are perceived to have vastly different levels of well-being, then they violate not only a sense of egalitarian parity but also a Rawlsian sense of distributive justice. However, if we start with two equally well-off societies and apply the same discount rate to each to gauge the effect of a particular GCC-mitigating investment, and we thereby estimate that one society will become much better off than the other in the future, this procedure would fail the Rawlsian measure of consequential equity in an intertemporal sense.

Only the "proportionality" measure for comparing outcomes implies an ethical need to link cause and effect, and therefore, it implies an explicit dynamic sequence in making those comparisons by associating "burdens or benefits—in proportion to the contribution of claimants" (IPCC 1996c, 86). Not only does this "proportionality" notion of equity align itself with the economist's prescription for efficiency—which results from the distribution of output in competitive markets (factors of production are paid the value of their marginal product), it may also coincide with the philosopher's notion of "fault-based equity" (Shue 1993, 10) used in tort law and as the basis for the "polluter pays" principle (IPCC 1996c, 105). So, while not explicitly dynamic, a comparison of outcomes (consequential equity) over time can trigger subsequent distributional actions. Because of the difficulty of assigning intertemporal welfare weights in the context of forging GCC-related policies, and because we claim that applying the discount rate is not a valid method for doing so, we must search for insights through these comparative notions of consequential equity and fault (action-triggered) distribution. Unfortunately, the IPCC report separates the discussion of equity principles in general (in its Chapter 3) from the discussion of discounting (in its Chapter 4). We seek to reunite them here.

In terms of "procedural" equity, the method of applying a discount rate to reach a decision that has intertemporal consequences might be said to satisfy that criterion if society finds it acceptable. However, the inability of members of subsequent generations to voice their opinions about that procedure (if they could, no doubt their views would rely heavily upon the estimated consequences of employing that mechanism) raises serious doubts about its satisfying procedural equity criteria. This example also emphasizes how notions of procedural and consequential equity frequently intertwine.

What is not generally viewed as procedural equity is the allocation that must often be made as a precondition for implementing a new state or process. Thus, even if a proposed change is perceived to be equitable, agreement to proceed may be difficult to attain unless the initial conditions (frequently unequal) are "grandfathered," thereby guaranteeing continued consequential inequity. Just the opposite would be true with respect to the Climate Convention if, as environmentalists encourage, many of the wealthier nations were willing to give up a greater amount of their current wealth by incurring current costs to forestall climate change, primarily to gain broad participation in an accord. Presumably, one reason for this voluntary transfer by rich nations, in addition to altruism (consequentialist equity), is the prospect that without an accord, as poorer nations continue to develop rapidly economically, future generations in those developing countries may swamp the contributions to GCC that the currently wealthy countries will make in the future (E_2^{AB} and E_2^{BA} in Figure 1). In this case, current generations of the wealthier countries would be willing to give up something in the hope of forestalling something far worse in the future. By comparing the current costs of reducing GCC with estimates of the future harm of not doing so, we can make inferences about intergenerational welfare weights (as assessed by the current generation). Furthermore, we can infer a discount-rate-like set of weights from such actions, but those weights probably have very little to do with traditional notions of the marginal SRTP.

Other Complications: Public Goods, Enforcement, and Sustainability

Reducing atmospheric concentration of greenhouse gasses, if it reduces the likelihood of GCC, will have varying impacts at different places and times on people who value those likely effects differently. So, although GCC reduction is a public good, its impacts are likely to be unequal since different climate effects will occur at different locations. Furthermore, even if the effects were identical, they might be valued differently by different individuals and societies. Therefore any action implies substantial explicit or implicit equity weights to account for the differential benefits imparted (IPCC 1996c, 110; Chilchilnsky and Heal 1994).

Consistent behavior across generations cannot be guaranteed. Unless we can commit subsequent generations to continuing the abatement activities begun by current generations (and unless, of course, the whole effort subsequently proves to be useless), posterity may negate the actions by current generations. However, since most societies currently deem human bondage contracts illegal, the fact that we cannot bind future generations to carry out our current wishes should be less troubling on ethical grounds. So, even if the dominant current ethical criterion is to pursue the biological imperative (sustainability), taken together these two principles imply that while our current actions should not doom future societies to extinction, it may be perfectly all right for future societies to choose to establish different values and not worry about their posterity, even though it is also ours. Once we act, posterity is in the driver's seat unless our actions eliminate posterity. We may expect members of different generations to hold different values. As an example, how many children and grandchildren have been disinherited because of their "different" lifestyles? But how can present generations take back an inheritance from a posterity based upon undesirable behavior they never observe? With free choice and no intertemporal enforcer of values, contradictory behavior across generations is certainly possible.

Two requirements for sustainability compound this dilemma, particularly for poor societies. Certainly we must leave a sufficient stock of biological support, knowledge, and physical capital (all infrastructure stocks) to posterity so future generations can do as well as current generations. But current generations must also survive if there is to be a future generation. As an extreme example, extinction today guarantees extinction in the future; the reverse is not necessarily true. So, in terms of implicit discount rates, saving for posterity implies a low discount rate. But if much of nature must be exploited today in order to guarantee that current generations will survive until tomorrow—a prerequisite for guaranteeing the existence of a future generation to worry about—then a relatively high discount rate today may be essential for survival. In practice, a proper market-related rate may take care of this second problem since it implies members of the current generation trying to arrange asset development and consumption among themselves over their lifetime so as to enhance their chances of success. In periods of great risk, one also might expect saving for a rainy day (but not for posterity) to be the type of behavior that enhances the chance of sustainability. This is different behavior than "eat, drink and be merry for tomorrow we might die," which gives no consideration to sustainability!

Ecologists in particular like to emphasize that reining in the earth's current population growth is a second important way to guarantee the survival of future generations, leaving adequate resources available so as to sustain humankind. In fact, nature performs this function in most cases naturally. But for humans, the population control route to near-term sustainability also poses difficult equity questions. In its severest form, in the face of crushing resource shortages, the question of intertemporal ethics is transformed into a question of intragenerational (contemporaneous) interpersonal welfare weights. Who alive today is to perish so that some survive so there will be a next generation? Obviously this is not a question currently faced by developed nations, but it is at least an implicit choice faced regularly by struggling impoverished societies. Frequently, only procedural equity (if any thought at all is given to equity in these circumstances) precepts are followed by the majority, and those with the greatest initial capacity are the ones most likely to survive. The major point to be picked up later is that *questions of intergenerational equity have implications for intragenerational equity*, and vice versa. Furthermore, to the extent that a discount rate is used as a surrogate for intergenerational welfare weights, then the choice of that rate may also have current spatial effects and, therefore, contemporaneous equity implications.

More Problems: Indivisibilities and Market-Related Discount Rates

If as Scarf has reminded us, in the face of indivisibilities (which surely abound in the issues at hand, such as scale economics, threshold effects, and extinction) the welfare theorems of economics don't necessarily hold, then extreme caution should be exercised in using market-determined prices and interest rates that arise from sectors in which indivisibilities are pervasive to make welfare inferences. Since the entire question of forging GCC-related policies is to forestall the possibility of extinction by ensuring at least minimal adequate resources for human survival in all societies (two examples of indivisibilities in nature's production and in consumption), then no single discount rate is necessarily required to project a desirable range of actions on future generations. Furthermore, doing so may be impossible because of an inability to enforce such arrangements.

The problems created by indivisibilities and by societies' response to them (frequently in the form of market adjustments) are extremely complicated and lead to substantial distortions of prices away from marginal costs, particularly within organizations that must balance their budgets (Schuler 1993). These considerations cast even more doubt on the desirability of searching for some unique market-related, social rate of discount to be used in formulating GCC policies, particularly if those market rates have been distorted by society's application of welfare weights for a purpose far different than gauging intergenerational equity.

PRACTICAL IMPLICATIONS FOR GLOBAL CLIMATE-CHANGE POLICIES

Equity and the Discount Rate: Their Use in the Policy Process

Any policy process usually comprises four functional parts: planning, authorization, implementation, and evaluation. Typically economic analysis plays the greatest role in the first and last functions, particularly through cost-benefit analyses. Public-choice theory focuses on authorization, and the "new institutional economics" has begun, in an analytic way, to pay attention to implementation. And on a practical basis, public administration has always attempted to deal coherently with all four functions.

Traditionally, however, cost-benefit analyses have been the principle area of economic concern, relying heavily on selection of the proper social rate of discount to weigh costs and benefits that are distributed widely over time. When equity has been considered, if at all, it generally has been in a retrospective sense of gauging the winners and losers of a particular policy. Over the past decade, however, prior estimates of the distributional consequences, particularly for policies related to economic development, have become an integral part of the planning process. No doubt driven in part by the perceived inequitable outcomes of many historic development programs, the prior estimation of distributional consequences has become a necessary prerequisite to gaining political approval. Thus, equity considerations have entered the planning process in order to improve the chances of the authorization of a particular policy. And while public administration practitioners may have dealt with process equity in a pragmatic way, and political scientists in a theoretical way, only recently have economists begun to explore in detail the necessary institutional components of dynamic processes, with respect to both equity and efficiency.

What is fascinating is that—although the importance of the discount rate in assigning proper weights to both outcomes and costs of dynamic processes distributed over time has long been recognized—only recently has equivalent attention begun to be paid to the equity implications that usually determine the dynamics of the process (including whether or not there is one).

Equity and the Discount Rate: Tests of GCC Objectives and Policies

Sustainability. The recurring theme alluded to in support of the Framework Convention is sustainability (IPCC 1996c, 90), as outlined in the 1987 Bruntland Commission Report. All of the many varying notions of sustainability imply building and/or leaving a *sufficient* stock of *essential* assets to ensure posterity has a *reasonable* chance of achieving that objective. Much of the debate centers on the relative importance of the italicized adjectives and on the components of and means taken to satisfy those concerns.

Another broader taxonomy divides those essential assets into physical, human, and natural (biological) capital. Some combination of all three is required to maintain current standards of living, and each has the similar characteristic that foregone current consumption and leisure can be invested to increase the stocks of these assets. Similarly, without some annual maintenance or reinvestment, the stocks will gradually erode. This is particularly true of physical capital; although, the ability to use human knowledge is also depleted over time if subsequent generations are not educated about the existence and use of that knowledge. However, if totally depleted, both human and physical capital can be re-created. In this last respect biological capital is unique: extinction is irreversible. Furthermore, without predators, nature normally sows the seeds of renewal for biological resources. Only when predators deplete the asset sufficiently such that an inadequate stock remains to guarantee renewal does a conscious savings effort have to be instituted (as is being enacted currently for New England fisheries).

Capital markets are used predominantly to determine investment in physical capital. Therefore market interest rates reflect, approximately, human intertemporal preferences for this type of asset, at least over individual lifetimes. While many commodity markets exchange the outputs from biological capital, the relevant time horizons are at most a season or two: given the possibility of extinction (an indivisibility), it is difficult to precisely infer from these markets longer term human values with respect to biological assets. Finally, if a wide variety of contracts for human bondage were permitted, the markets for human and physical capital could merge, and equally valid inferences about intertemporal preferences might be drawn. However, this condition is not met in most societies; that is why subjective welfare judgments usually weigh heavily in public policies on education and on research and development.

Still, market rates of interest, dominated in their determination by markets for physical capital, can tell us the opportunity cost of investments in human or biological capital. They tell us what we could have reasonably expected to have gotten for our money if, instead of investing in biodiversity or in developing new knowledge for posterity, we had allocated it to physical capital.

Since the intended beneficiary of GCC policies is posterity, and since there is no guarantee that any action taken now will be delivered to or utilized by that posterity in the way present generations intended, the best that current generations can do is establish and maintain the necessary stocks of assets that form prerequisites for sustainability and continue the formation and trial use of international organizations and mechanisms that might deal with the problem as its severity becomes more apparently widespread. Of all these stocks, information may be the most valuable and durable. Investments in human capital can certainly improve our understanding of the sources, sinks, and atmospheric balances of greenhouse gases and of the relationship between those concentrations and GCC. R&D that improves the resource efficiency of human activity, including reduced population pressures, may also be of long-lasting value. And institution-building can increase the speed of response to future events and reduce adjustment costs appreciably. Furthermore, this approach may be consistent with modern options theory for investments in an uncertain environment.

While any decision about the level of current actions needed to offset potential future harm suggests an implied discount rate, the dominant rationale given for recommendations is contemporaneous equity among nations (E_1^{BA} and E_1^{AB} in Figure 1). Who is to do what, now, and who will be the short-run beneficiaries (as well as the long-run potential gainers)? The other current allocation decision is based upon assessments of risk and largely involves decisions about what insurance stocks of these assets that are essential for sustainability should be accumulated now. Again, the level of these investments needed to hedge against catastrophic risk is determined in part by an implied discount rate, but since both the action taken and the resulting economic activity will be widely scattered geographically, contemporaneous equity assessments will likely dominate the decisions.

IPCC Principles and Actions. In fact, the IPCC's "guiding principles" do just that: they spell out who the current winners and losers ought to be in terms of contemporaneous (international) equity, (IPCC 1996c, Section 3.5). The primary allusion to fault-based equity is to insure parties in accordance to their varying needs, which acknowledges that GCC is likely to affect some nations more severely than others. A primary justification—given for what is in fact a contemporaneous redistribution under the label *"differential responsibility"*—is that since rich nations have contributed the most to existing atmospheric concentrations of greenhouse gases, they should be responsible for the greatest cleanup. While alluding to a fault-based, polluter-pays principle of equity, this is purely a contemporaneous equity allocation since the Framework Convention does not spell out future consequences in response to future emissions. As an example, efficient future remediation might well imply that if less-developed nations using even dirtier methods of production than the rich nations found it less costly to engage in a unit of cleanup than the rich countries, then an appreciable amount of the future remediation should take place in the poorer, developing nations.

Actions implemented jointly (AIJ) is a process of allowing voluntary bilateral arrangements that attempt to achieve these efficiencies in remediation, given the agreed-upon, equity-driven initial allocation of responsibilities.

The *global environmental facility* (GEF) is another mechanism for implementing contemporaneous redistributions that has been accepted, in principle, by the ratifiers of the Framework Convention. In this case the wealthy nations contribute funds to the facility, which in turn supports worthy projects in the developing nations. Thus the redistribution is into a common pool of funds, and the subsequent allocations are made administratively with a focus on international equity and on spawning learning and innovation in the developing countries. The GEF has come under heavy attack by many developing nations, however, based upon process-equity concerns because the initial administration of the fund relies heavily on the World Bank and other international organizations in which developed nations have a strong voice.

The problem with this mèlange of policy instruments is that it intertwines elements of contemporaneous, consequential equity with fault-based and procedural equity in what may be a pragmatic assortment, but it makes evaluation of future cost-effectiveness, not only of the outcomes but also of the processes themselves, very difficult. Although the principles and goals of the Framework Convention are subject to review and revision, since all of the actions except AIJ are implemented administratively, those adjustments are likely to occur only after lengthy and prolonged debates; and much of the rationale and potential consequences may not be particularly transparent. AIJ does afford one mechanism to check and improve the efficiency of administratively imposed requirements as the number of AIJ opportunities increases beyond the demonstration phase. But if environmental constraints are tightened dramatically in the future, then the aggregate administrative costs of searching for and concluding these bilateral contracts may itself inhibit further efficiency gains.

In all of these GCC policy revisions, if they are administrative in nature, debates about the implied SRTP will continue to dominate with very little additional information (based on actual long-term negotiated settlements) provided to update these implied rates, other than through AIJ. That is the major problem with top-down, administrative-type regulatory processes that establish quantity-based quotas and restrictions: little updated information is received, except as measured by the degree of noncompliance and the number of participant complaints. However, when those quotas (or permits) can be traded, valuable additional information about their perceived worth is obtained to guide future policy-making. And when those trades transpire over time, some inferences about intergenerational values might be made, subject to all the previous caveats because future generations are not parties to these trades.

Illustration of a Transparent Alternative: An International System of Tradable Permits

Just as the Framework Convention required a general sense of fairness in order to gain its widespread ratification, so too a system of tradable permits must, at a minimum, satisfy the parties' sense of international consequential equity through the initial allocation of permits, of procedural equity and of their concerns for long-term consequential equity. With a system of tradable permits, however, these equity allocations are totally transparent (and therefore might enlighten subsequent negotiations). Furthermore, once in place, a well-designed system of tradable permits yields important updated information on compliance costs and perceived benefits over time.

Several authors (S. Barett, M. Grubb, A. Rose, R. Sandor, and T. Tietenberg) surveyed the possible dimensions of such systems, including equity approximations, in *Combating Global Warming* (U.N. 1992). Furthermore, R. Schuler (1994) describes a mechanism proposed by Heller, Rubenstein, and Schuler that makes the contemporaneous international equity implications explicit and provides enormous subsequent flexibility and national autonomy, while still maintaining transparency.

A key ingredient of these proposals is devising an initial allocation of permits that encourages existing and potentially significant future emitters of greenhouse gases to sign onto the agreement, thereby implicitly demonstrating that the agreement is deemed to be equitable. Once set, this initial stock of permits allocated to each nation would not be adjusted for subsequent population or economic development changes (underlying sources of GCC). However, the number of outstanding permits could be adjusted (phased down) gradually over time, consistent with adequate protection for the environment. Countries would be free to buy and sell permits, including through a central bank, and to trade against future allocations. If the permits issued to each individual country were "risk-rated" by potential buyers according to the selling nation's perceived future com-

Edited by R. T. Watson, M. C. Zinyowera, and R. H. Moss. Cambridge: Cambridge University Press.

———. 1996c. *Climate Change 1995: Economic and Social Issues of Climate Change. The Contribution of Working Group III to the Second Assessment Report of the Intergovernmental Panel on Climate Change.* Edited by J. P. Bruce, H. Lee, and E. F. Haites. Cambridge: Cambridge University Press.

Lind, R. C. 1990. Reassessing the Government's Discount Rate Policy in Light of New Theory and Data in a World Economy with a High Degree of Capital Mobility. *Journal of Environmental Economics and Management* 18(2): S20–21.

———. 1995. Intergenerational Equity, Discounting, and the Role of Cost-Benefit Analysis in Evaluating Global Climate Policy. *Energy Policy* 23(4/5): 379–89.

Lind, R. C. and others. 1982. *Discounting for Time and Risk in Energy Policy.* Washington, D.C.: Resources for the Future.

Loewenstein, G. and R. H. Thaler. 1987. Intertemporal Choice. *Journal of Economic Perspectives* 3(4):181–93.

Manne, A. S. 1995. The Rate of Time Preference: Implications for the Greenhouse Debate. *Energy Policy* 23(4/5): 391–94.

Munasinghe, M. 1993. *Environmental Economics and Sustainable Development.* Washington, D.C.: World Bank.

Nordhaus, W. D. 1994. *Managing the Global Commons: The Economics of Climate Change.* Cambridge, Massachusetts.: MIT Press.

Ramsey, F. P. 1928. A Mathematical Theory of Saving. *Economic Journal* 138(152): 543–59.

Rawls, J. 1971. *A Theory of Justice.* Cambridge, Massachusetts.: Harvard University Press.

Scarf, H. E. 1994. The Allocation of Resources in the Face of Indivisibilities. *Journal of Economic Perspectives* 8(4): 111–28.

Schelling, T. C. 1995. Intergenerational Discounting. *Energy Policy* 23(4/5): 395–401.

Schuler, R. E. 1993. *Pricing over Space and Time with Fixed Costs and Stock Externalities.* Center for Analytic Economics. Working Paper 93-18. November. Ithaca: Cornell University.

———.1994. *International Mechanisms to Achieve Voluntary Reductions in Atmospheric Concentrations of Greenhouse Gases.* January. Paper presented at ASSA Annual Meetings, Boston, Massachusetts.

Shue, H. 1992. The Unavoidability of Justice. In *The International Politics of the Environment: Actors, Interests and Institutions,* edited by A. Hurrel and B. Kingsbury. Oxford: Clarenden Press.

———. 1993. *Environmental Change and the Varieties of Justice: An Agenda for Normative and Political Analysis.* Paper presented at a conference on Global Environmental Change and Social Justice. Carleton and Cornell Universities, September 10–12.

U.N. (United Nations). 1992. *Combating Global Warming: Study on a Global System of Tradeable Carbon Permits.* New York: UNCTAD/RDP/DFP/I.

Comments

Equity and Discounting in Climate-Change Decisions

William R. Cline

Robert Lind and Richard Schuler deserve congratulations for their stimulating and lucid review and their attempt to identify a new approach to intergenerational discounting. However, in their questioning they would seem to throw away too much. In effect, they say "a plague on both your houses" to the two sides in the prescriptionist-descriptionist debate. In doing so, they retreat to nonquantifiability. This effectively concedes what scientists have said all along: that economists cannot determine whether abatement is warranted. In my view, it is a mistake to throw in the towel in this way, and the two authors' despair that there is "no discount rate" is unjustified.

Lind and Schuler are exactly right that the "fatal flaw" of the descriptionist approach is that it is impossible to undertake an ironclad commitment to invest over two hundred years. However, the authors are wrong to reject the prescriptionist view. They do so primarily on the basis of the Schelling critique. Yet this critique is a red herring at best, and more likely even goes in the wrong direction logically for rejecting the *social rate of time preference* (SRTP). Thus, I conclude that the prescriptive approach should be retained; reject the Schelling argument; and, agreeing with Lind and Schuler, also reject discounting over centuries at today's rate of return on capital.

The authors of the discounting chapter (Chapter 4 of the IPCC report) made every effort to reflect the diversity of views among economists examining climate change and, in doing so, may have left undue

WILLIAM R. CLINE is Deputy Managing Director and Chief Economist at the Institute of International Finance, Washington D.C.

ambiguity about their skepticism on the position of the descriptionist school. The chapter authors broadly supported the prescriptive approach and, in an initial draft, included a table with illustrative parameters and discount rates based on this approach. In the final draft, however, the authors' critique of the approach of discounting at today's rate of return on capital probably remained too implicit.

I for one was somewhat uncomfortable with the IPCC chapter's very use of the "descriptive-prescriptive" nomenclature. In my view the empirical, and thus descriptive, evidence shows that the real rate of return at which consumers can transfer consumption into the future is the risk-free real rate on treasury bills, which historically has been close to zero. There is thus a descriptive basis for saying the rate of pure time preference is zero. Nor can we "describe" what the rate of return on capital will be over the next two or three centuries, let alone whether society can be counted upon to set aside extra resources at this rate to compensate for future greenhouse damage. "Prescriptive" sounds arbitrary and personal, yet the SRTP approach is just as empirically based as the alternative capital rate of return approach.

Fortunately, Lind and Schuler sharpen what the IPCC chapter said sotto voce:

> [W]e don't know how to go about setting up such a trust fund. In the case of global warming, we would somehow be required to set aside special resources on a worldwide basis and continue to invest and reinvest them at the going market rate of return until the time that compensation was to be paid. In addition to that problem, there are the problems of making sure on an international basis that these resources would go to the people in the future who would actually incur the damages, and of making sure that the value of these resources was not diminished by global warming itself.

The closest the IPCC authors came to stating this skepticism was in a box giving a numerical example, where they noted that the option of "[o]ther investment...may be institutionally infeasible, as there may be no way to put aside $1 million today and leave it untouched for 50 years as a Fund for Future Greenhouse Victims" (Arrow and others 1996, 134). They also added that "the descriptive approach assumes compensation from one generation to another for any loss of environmental amenities, implicitly leaving unanswered whether compensation is likely to occur" (133).

Lind and Schuler nicely point out that intermediate generations can hijack capital stocks intended to compensate distant future generations. The fiscal deficits of the Reagan era provide a good example.

The two authors make another important criticism, almost in passing: that it is a grievous mistake to take present infinitesimal marginal rates as applicable over centuries. This coincides with my own critique that it is unlikely that today's rate of return on capital can persist indefinitely with capital deepening, even allowing for technical change.[1]

I disagree with the Lind-Schuler critique of the SRTP approach. By way of background, during the course of the IPCC Working Group III meetings, it became clear that several of the chapters kept coming back to the same formula for the discount rate, explicitly or implicitly:

$$\text{SRTP} = \rho + \theta g$$

where ρ is the rate of pure time preference (impatience), θ is the elasticity of marginal utility, and g is the growth rate of per capita consumption.

I suggested that a separate chapter be added on discounting to consolidate this central stream of analysis. Two of the authors of the resulting chapter (Kenneth Arrow and Joseph Stiglitz) had made earlier seminal contributions on the social rate of time preference, and naturally they saw this approach as appropriate for analyzing global warming.

The Lind-Schuler critique of the SRTP approach is equivalent to the critique of Thomas Schelling. Essentially they argue that society does not apply the same value judgment implicit in the SRTP to intercountry, intratemporal transfers today, so there is no reason to accept this utility function for intergenerational transfers. There are several problems with this critique.

First, consider a 2×2 matrix, with 1996 and 2200 as the two columns and North and South as the two rows. Just because there are limited flows from North to South in the first column, despite low per capita income in the South, it does not mean that there is a lack of concern along each of the rows: by Americans today for Americans in 2200, or by Bangladeshis today for Bangladeshis in 2200.

Second, the critique confuses individuals with society. Any individual American has the right to disinherit his unborn descendants because he holds the same empathic distance from them as he does from present day Bangladeshis. But U.S. society as a whole is not similarly free to be oblivious about unborn U.S. descendants. If President Clinton announced tomorrow that he has no concern for America two hundred years from now, voters would react negatively. Instead, he has just tightened his greenhouse abatement commitment in an election year.

Consider a thought experiment. Proposal A sets aside one-third of the state of Utah as a national park for preservation for our descendants two hundred years from now. Proposal B is to grant one-third of Utah to Bangladeshi immigrants for strip mining. A candidate announcing proposal A would get many votes; a candidate announcing proposal B would be considered mentally unbalanced. Yet the Schelling (and Lind-Schuler) argument requires that the public should be indifferent between the two proposals. In short, the public does feel a commitment to the future of its *own* descendants, even though they are unborn, and even though it feels little if any commitment to citizens of other nations.

Third, efficiency is a reason for limits on actual intragenerational transfers. The U.S. marginal tax rate was once above 70%, and that in the U.K. over 90%. These high rates were reduced because of their inefficient incentive effects, approximating Laffer-curve results. The fact that strongly progressive rates were tried, however, and that some progressivity remains indicates the public's welfare function is not wholly unlike that underlying the SRTP.

Fourth, the Schelling argument can, in fact, be turned on its head and lead to support rather than rejection of the SRTP. Fundamentally, the critique means that the elasticity of marginal utility is close to zero. That is, if we do not see transfers from countries with $25,000 per capita income to countries with $1,000 per capita income, then implicitly the marginal utility of a dollar's consumption is identical between the two, and the elasticity of marginal utility (θ) is zero.

But this goes in the wrong direction for rejecting the SRTP as yielding too low a discount rate. Consider the following syllogism. (A) The Schelling critique implies that $\theta \to 0$, rather than > 1 as suggested by the prescriptionists. (B) A lower θ reduces rather than increases the SRTP. (C) So the Schelling argument cannot be used to contend that the SRTP yields too low a discount rate. Correspondingly, it cannot be invoked in support of the descriptionist position calling for a higher discount rate.

Fifth, Lind and Schuler, and Schelling make the mistake of equating greenhouse gas mitigation decisions to the bestowal of a benefit, whether to Bangladesh today or our own unknown descendants in the future. Instead, the issue is the imposition of a damage. Surely there is an ethical difference between refraining from conveying a gift, on the one hand, and imposing a damage, on the other. Americans might feel no compelling obligation to increase aid to Bangladesh today, but surely they would be loathe to despoil Bangladesh today (for example, by holding nuclear tests close by offshore).

Sixth, another aspect of the Schelling critique is the proposition that no particular reason exists for postulating exponential decay in discounting. Of course, one can easily envision some sort of step function with lower but horizontal steps at more distant generations, perhaps with unborn generations all given identical consideration rather than facing exponential decay in the order of their appearance (Rothenberg 1993). The problem is that Schelling would seem to place the level of the final step at zero, whereas there is no reason why the final step might not be much higher.

In sum, Lind and Schuler have unfortunately been captivated by the argument that it is impossible to evaluate intergenerational transfers because we do not make intercountry, intragenerational transfers according to the same utility function in other words, the Schelling critique. Yet, for the reasons just set forth, this critique is a red herring.

The unfortunate consequence is that rejection of the SRTP approach on the Schelling grounds leaves Lind and Schuler at sea, with no answer to the discounting problem. They propose instead some time graphs for policymakers' elucidation. But the policymakers will not know what to make of these graphs if the economists are incapable of saying what they mean.

My time graph of the greenhouse issue shows the costs of abatement greater than the benefits of avoided damages early in the time horizon, but abatement costs far below benefits later in the horizon. With no judgment on discounting, it is impossible to make an overall evaluation of proper abatement.

The Lind-Schuler conclusion would basically return economic analysis of global warming back to agnosticism about the value of mitigation and limit economists to the role of quantifying abatement costs if politicians decide independently (presumably on scientists' advice) to undertake abatement. To my taste this is unsatisfying.

The Lind-Schuler conclusions are right to emphasize that we cannot be sure that per capita growth will average as high as 1% to 1½%, with the result that per capita incomes will rise some thirtyfold over the relevant time horizon. The public does not believe such a brilliant future lies in store.

They are also right to emphasize that we cannot rule out a catastrophe. My own analysis (Cline 1992) finds, using the SRTP approach with some weighting for risk aversion, that abatement is warranted even without taking account of catastrophes. Their inclusion would simply reinforce this policy finding. As a footnote, however, the Lind-Schuler paper implies that global warming could lead to extinction. Yet, while extinctions of some species will surely occur, I have never seen even a catastrophic analysis that suggests Homo sapiens will be among them.

I also agree with the paper's emphasis on the importance of sequential decisionmaking. A separate IPCC chapter addresses uncertainty and risk. But again the implied Lind-Schuler approach is inadequate: it would look out over a horizon of say thirty years and discount normally over this period to evaluate abatement costs. The problem is that this approach would provide no idea whatsoever about the benefits of abatement.

Finally, a further word about the prescriptionist versus descriptionist schools. Lind and Schuler's equation (1) shows the rate of return on capital i as equal to the consumer rate of discount r and both as equal to the SRTP. However, there is a "wedge" of tax and other distortions between the return to capital on the one hand and the consumer rate of discount and the SRTP on the other. We should then instead write:

$$i[1 - \psi] = [\rho + \theta g]$$

where ψ is the proportion of leakage from capital return to tax and other distortions. Chapter 4 of the IPCC report notes the use of $\rho = 0$, $\theta = 1.5$, and $g = 1\%$ in Cline (1992), leading to SRTP $= 1\frac{1}{2}\%$; and suggests an upper limit of 3%. It is not difficult to square an SRTP of 3% with observed market rates. With $i = [\rho + \theta g]/[1 - \psi]$, a leakage rate of $\psi = 0.5$ would mean that a marginal capital return of 6% is consistent with consumer and SRTP rates of 3%. Marginal tax rates alone could make ψ as high as 0.5. So then we may ask whether 6% is a reasonable rate for marginal capital return.

At the macroeconomic level, capital share in industrial economies is about 20% of national income. With a capital/output ratio of about 3, which recent international estimates tend to reaffirm, and with the net rate of return on capital equal to net capital earnings divided by capital stock, we have:

$$i = 0.2Y / 3Y = 6.7\%$$

where Y is national income. As this is the average rate, the marginal rate is presumably lower. So at least the 3% ceiling for the SRTP *is* descriptive in the sense of being consistent with what we observe. The relevant issue then becomes whether to discount at the gross marginal capital return rate or at the consumer rate net of the wedge. The SRTP approach does the latter (after converting capital effects to consumption equivalents using a shadow price of capital). Because Lind and Schuler rightly reject the notion that we can commit to a program of investment over centuries, they strike a devastating blow against using the capital rate of return. They strike at most a glancing blow, however, against the SRTP approach.

ENDNOTE

1. Some of the descriptionist-school models, however, incorporate declining capital return over time. Thus, the DICE model (Nordhaus 1994) optimizes discounted logarithmic utility with a production function that has capital deepening and technical change declining. This approach yields a declining growth rate and marginal product of capital and, hence, what is equivalent to a falling discount

rate component $\theta\gamma$ with $\theta = 1$ in the logarithmic function. My quarrel with the model's discounting is instead with its use of a non-zero pure time preference: $\rho = 3\%$.

REFERENCES

Arrow, K. J., W. R. Cline, K.-G. Mäler, M. Munasinghe, R. Squitieri, and J. E. Stiglitz. 1996. Intertemporal Equity, Discounting, and Economic Efficiency. In *Climate Change 1995: Economic and Social Dimensions of Climate Change*, edited by J. P. Bruce, H. Lee, and E. F. Haites. Cambridge: Cambridge University Press.

Cline, William R. 1992. *The Economics of Global Warming*. Washington, D.C.: Institute for International Economics.

Nordhaus, William D. 1994. *Managing the Global Commons: The Economics of Climate Change*. Cambridge, Massachusetts: MIT Press.

Rothenberg, Jerome. 1993. Economic Perspectives on Time Comparisons. In *Global Accord: Environmental Challenges and International Responses*, edited by C. Nazli. Cambridge, Massachusetts: MIT Press.

Comments

Equity and Discounting
in Climate-Change Decisions

Richard N. Cooper

This comment concerns how we today should weigh actions that will have their main beneficial effects in the future, possibly (as with mitigating potential climate change) in the far distant future. Economists have long argued that in comparing costs and benefits that are spread over time, with costs and benefits not generally matched, the costs incurred or the benefits enjoyed in the future must be "discounted," with a greater discount applied to the more distant future, to achieve a useful net present value.

The very idea of discounting is controversial in some circles because it is mistakenly assumed to imply that we accord less weight to the well-being of future generations. (See for example Cowen and Parfit 1992). Yet not to discount the future leads to absurd conclusions: any finite investment that yields a finite return indefinitely, however small, should be undertaken, since the cumulative return will always exceed the investment. Thus today's generation should cut its consumption to subsistence levels for the sake of all future generations. This defies common sense. (This concept underlay Soviet investment in the 1950s, leading to excessively capital-intensive projects until Russian engineers discovered their mistake.) Common sense is not always valid, but when we give advice that runs counter to common sense, there should be compelling reasons to support the counter-intuitive advice.

The IPCC authors acknowledge the need for discounting, and without being too technical, Chapter 4 contains what in many respects is a

RICHARD N. COOPER is Boas Professor of International Economics at Harvard University.

sophisticated discussion of the reasoning behind discounting (IPCC 1996). Once the need to discount is granted, a discount rate (or a sequence of them) must be chosen. On this issue, unfortunately, the chapter is fundamentally misleading. It describes two competing schools of thought: the prescriptive (or ethical) school and the descriptive (unethical?) school. The prescriptive approach "begins with a social welfare function constructed from ethical principles" (131). The descriptive approach "focuses on the (risk-adjusted) opportunity cost of capital" (132). But the major dispute is actually between those who look *prescriptively* at the opportunity cost of capital, and those who attempt to construct an abstract, academic social-welfare function, which typically assumes that many critical features of the (implied) economy are in equilibrium. *Either* approach then requires some principle for "description" in order to generate an actual number (or numbers) for the discount rate. The opportunity-cost school typically looks at various market-based empirical observations, possibly adjusted in some way; the social-welfare-function school typically looks to historical observations (for example, growth in per capita incomes), to experimental observations (for example, the bounds of the hypothetical marginal utility of consumption), and to either contemporary empirical observations or pure *a priori* reasoning (that is, personal preference) to quantify the pure rate of time preference.

In my view, the opportunity-cost school unambiguously dominates the social-welfare-function school as a *prescriptive* matter. That leaves open, in both cases, how one finds actual numbers, although I am influenced by the thought that numbers are needed to be operational. The so-called prescriptive school uses highly arbitrary numbers (as well as an arbitrary formula for its social-welfare function). Without actually saying so, the IPCC authors implicitly suggest that the prescriptive school is preferable to the descriptive or opportunity-cost school, especially if there is no way to assure that the right people (as specified in the social-welfare function) get the benefits of an investment undertaken today.

The main reason for my preference is straightforward: so long as present consumption and future consumption both have value to individuals and to society, a social loss is involved if future consumption is not maximized, given current consumption; put another way, a social loss is involved if current consumption is not maximized, given future consumption. This point seems self-evident, assuming we can measure opportunity cost adequately (the "descriptive" part). In other words, we do future generations a *dis*service, not a favor, if we pass up known high (social) return investments for lower return ones: yet this is what we will do if we choose too low a discount rate for evaluating climate-change mitigation actions and then actually undertake them. It is no response to point out that some people save at very low rates of return (a descriptive point). If we can extract funds from such people cheaply, fine; but we still serve future generations best, and at least cost to the current generation, by investing those funds in high-return activities rather than low-return ones.

This sounds like common sense. What can be the objection to it? One possibility is this: in the near future the return on investment class B (mitigation actions) falls short of the return on investment class A (some other investments). However, in the long run the reverse is true because of a secular decline in returns to investment class A. Normally, we could switch to B investments as returns on A drop below those on B. The preference for investing in B now would occur only if for some reason it would be too late to switch to B investments later, after returns to new A investments fell.

This type of configuration is theoretically possible, but it is necessary to make a plausible case for both parts to conclude that we should reject A in favor of B at the outset. In the standard neo-classical economic model, the returns to capital are assumed to fall steadily as the ratio of capital to labor (and other factors) rises. But in historical—as distinguished from analytical—time, technological change has constantly increased the returns to (new) capital, and there is no reason to believe that that process will stop during the next century. Thus if returns to class A are high now relative to B, they are likely to remain so.

The IPCC authors seem to reject the efficiency argument that is emphasized here, not on the foregoing grounds, but on the basis of equity. We cannot ethically say that investment A is superior to investment B—even if it yields higher total future benefits—if those persons who experience losses as a result of the investment are not actually compensated (in the absence of a social-welfare function that indicates the relative weights we should attach to prospective winners and losers).

This point is logically valid. But if taken literally and applied seriously, it is a prescription for total inaction, especially when time frames as long as one hundred years are under consideration. Again, it defies common sense.

First, we have not collectively agreed on a social-welfare function, and we have no prospect of agreeing on one at a global level. So we cannot generally weigh winners against losers, especially over so long a time period.

Second, we cannot possibly know future winners and losers from our actions today (try, for example, to forecast reasonably accurately the winners and losers from completion of the U.S. continental railroad in 1869, or the winners and losers one hundred years hence from construction of the new Three Gorges Dam in China.)

Third, we could not bind future generations to adhere to our preferred outcomes even if we could have the requisite knowledge about future winners and losers and our preferences among them. If we make rules, future generations can unmake them. If we plant trees, they can cut them down. If we consume less coal, they can consume more—and may actually do so because it is more readily available to them. The one legacy we can leave that is impossible to reverse (short of a collapse of civilization) is enhanced knowledge, both a deeper understanding of nature and improved technology.

We should be concerned above all with passing to the next generation more knowledge and higher incomes than we received, and allow its members to decide how to distribute them—they will so in any case, regardless of what we think.

This is not to suggest we should be completely indifferent to distributional effects. Our actions will affect the initial distribution of the next generation, and collectively we may want to avoid certain actions on the grounds that we do not like their distributional effects. Here I mean the direct next-generation consequences of our actions, where it may be possible to agree collectively to avoid imposing extreme losses on certain classes of people. But we cannot carry this logic into the more distant future on the grounds already mentioned: we cannot possibly know the distant future impact on people (for one thing, we do not even know where they will be), and we cannot commit future generations to our preferences even if we did.

In any case, it is rather odd to urge costly action now for the sake of poor people in the distant future when we are not willing to take very costly action now for the sake of reducing poverty today. We have actual evidence on the amounts we are willing to spend, individually and collectively, in the name of reducing poverty in today's poor countries: about 0.3% of GDP of the rich countries. If we are really concerned about the impact of possible future climate change on poor people, we should take more active steps now to reduce their poverty systematically, which in principle we know how to do. That would improve their capacity to adapt to such climate changes as may take place, and to take mitigation actions themselves in the future.

So I conclude that, as a prescriptive matter, we should use the opportunity cost of other possible investments which benefit future generations to assess the merit of greenhouse gas mitigation actions today and in the near future. But what actual number is that? Evidence suggests that returns to education in developing countries exceed 20%. A study of more than one thousand projects completed by the World Bank in the 1970s and 1980s yielded an average (prospective) return of 16%. The World Bank and the U.S. government have stated threshold returns of 10% for evaluating prospective investments (recently reduced to 7% by the U.S. government). The corporate sector of the U.S. economy, one that is relatively rich in capital by global standards, yields an average pre-tax real return well over 10%. For all these reasons, I believe that 10% is a reasonable rate of discount. Maurice Scott of Oxford University has suggested 4% (as reported in Beckerman 1996), partly on grounds that that has been the real yield on low-risk government bonds in recent decades. But that would be a mistake: even if we can extract resources from the public at 4%, we should invest in those activities with high (social) return. Only after we exhaust 10% and 7% and 5% opportunities should we accept investments with prospective yields of only 4%. Otherwise we cheat future generations.

Some observers object to citing data on observed rates of return on grounds that actual decisions made today and in the past have not been made under ideal conditions and reflect a number of imperfections both in markets and in our processes for making collective decisions. It would take us too far afield to explore this contention in relevant detail. Let me just stipulate that by any given set of ideal standards, the real world is messy and actual decisions (and market outcomes) deviate from those ideals. What bearing does that have on the issue at hand? The same observation will apply to actions to mitigate greenhouse gas emissions. A plausible argument must be made that allowing for the various imperfections will raise the after-the-fact returns to mitigation actions relative to the observed returns on other investments.

The debate over the choice of a discount rate to be applied to mitigation actions can be interpreted as an effort to reduce or eliminate imperfections in collective decisionmaking on public expenditures in general. But if such imperfections are important, and if other public investments seem to leave future generations still better off, why don't the advocates of low discount rates first apply their arguments to those higher yield investments? Some of them no doubt do. Those who do not must fail to do so either because they believe the political prospects for improving collective decisionmaking are better in the arena of global climate change than for other, higher yield public investments; and/or they prefer mitigation of greenhouse gas emissions on some different and generally unstated grounds, not captured in the usual reckoning of costs and benefits over time, and they want to develop any argument that tends to support such actions. In either case, it would be useful to open these considerations explicitly to wider discussion.

REFERENCES

Beckerman, Wilfred. 1996. *Through Green Colored Glasses.* Washington, D.C.: CATO Institute.

Cowen, T. and D. Parfit. 1992. Against the Social Discount Rate. In *Justice Between Age Groups and Generations,* edited by P. Laslett and J. S. Fishkin. New Haven: Yale University Press.

IPCC (Intergovernmental Panel on Climate Change). 1996. *Climate Change 1995: Economic and Social Dimensions. The Contribution of Working Group III to the Second Assessment Report of the Intergovernmental Panel on Climate Change.* Edited by J. P. Bruce, H. Lee, and E. F. Haites. Cambridge: Cambridge University Press.

4

Applicability of Cost-Benefit Analysis to Climate Change

Paul R. Portney

INTRODUCTION

One of the distinguishing features of climate change is its complexity—along scientific, economic, political, and legal lines, among others. For that reason, anything that can help us think systematically about climate change, as well as proposed solutions to it, is most welcome. That is one of the virtues of cost-benefit analysis (CBA). Although a relatively minor branch of economics, CBA attracts much attention, particularly from those outside the economics profession, because it represents one way to try to weigh the pros and cons of a contemplated policy change—whether related to climate or not—and decide whether the former justify the latter.

It is not surprising at all, then, that those charged with producing a state-of-the-art document on the possible contributions of the social sciences to the climate-change debate elected to commission a chapter on cost-benefit analysis. Nor is it surprising that, in drawing their conclusions about what CBA can and cannot contribute to this debate, the authors' interpretations differed from the interpretations that many others might have drawn. The hope here is that a discussion of the inferences the authors drew, and how they compare with other possible interpretations, will be helpful to those interested in either CBA or the problem of climate change, or both.

PAUL R. PORTNEY is Senior Fellow and President, Resources for the Future.

In one sense, no one ought to be saddled with the job confronting the authors of Chapter 5, Applicability of Techniques of Cost-Benefit Analysis to Climate Change, of the report of Working Group III of the IPCC (IPCC 1996). After all, cost-benefit analysis has spawned a large and rapidly growing literature, with many of the recent contributions focusing narrowly on environmental and natural resource applications. In addition, as the authors of Chapter 5 point out, the application of CBA to decisions regarding the prevention and/or avoidance of possible harmful effects associated with carbon dioxide and other greenhouse gases is very, very difficult (though perhaps not so different from other problems as the authors assert). Therefore, any effort to treat the subject in thirty to forty pages is bound to frustrate many readers in at least some respects. This ought to be acknowledged at the outset.

For this reason, the right question to ask about the chapter, it would seem, is: How good a job did the authors do in explaining the usefulness of CBA to the question of climate change, given the breadth of the subject(s) and the limitations of space? My answer to this question is: a fair job. To be sure, they touched on many, if not all of the obvious problems that arise in using CBA to formulate response strategies. And they heavily emphasized what are probably the most problematical aspects of CBA in this application. In fact, Chapter 5 will make at least some economists think again about issues related to CBA that may have been more or less settled in their minds. It will also provide some useful information to those who bring to the Working Group III report little or no technical expertise in applied welfare economics.

Nevertheless, there is an uneven quality to Chapter 5 that is frustrating, and it contains at least some mischaracterizations of CBA or idiosyncratic views of modern policy analysis. Any of them taken alone could be excused, especially in a relatively short chapter on a complicated subject. Taken together, however, they detract from the useful points the authors make and create several misleading impressions about CBA and its potential utility. In addition, the authors of Chapter 5 erred in several more significant ways and missed an opportunity to address several fundamental issues concerning CBA and climate change that would have proven to be quite illuminating to decisionmakers.

The first section below provides a very brief overview of Chapter 5. Next, I have identified a number of "small-peanuts" issues that, when taken together, detract somewhat from the quality and utility of Chapter 5. Following that, I raise several issues that I feel should have been given far more attention than they were. Here, of course, I am offering my views on the difficulties that we encounter when applying CBA to real-world policy problems. I conclude with some recommendations for future work in this area.

OVERVIEW OF WGIII CHAPTER 5

It may be useful to briefly summarize Chapter 5 of the WGIII report. It begins with an overview of CBA. This overview distinguishes between CBA and other related and useful analytical techniques such as cost-effectiveness analysis, what the authors call multicriteria analysis, and decision analysis. As spelled out in more detail below, one criticism of Chapter 5 is that too little time is spent on CBA *per se*, and too much time is spent on alternatives to it. This is unfortunate because—given the title of the chapter—many readers will turn to it for a clear statement of the pros and cons of CBA in formulating climate policy.

In this context, I will begin with a definition of CBA so readers may judge for themselves whether the view laid out here is consistent with their own. Cost-benefit analysis is a technique used for policy evaluation in which all the favorable and unfavorable effects associated with a policy change are identified, quantified, and (whenever possible) evaluated in dollar terms. The time streams of benefits and costs are then generally discounted to the present so that a net present value can be calculated. When several possible policy interventions are being contemplated, CBA can be used to compare and rank them. The explicit effort in CBA to place dollar values on all benefits and costs is both a strength and a weakness, and is also that which, more than anything else, differentiates CBA from other analytical techniques

Following the discussion in Chapter 5 of CBA and alternatives to it, the authors devote a great deal of space to the special problems they feel climate change poses for CBA. These include the physical relationships that form the chain of causality, the "stock versus flows" characteristic of carbon dioxide accumulations in the atmosphere, the inertia and irreversibility that characterize the problem, its temporal and geographic reach, and the nonlinearities that may be inherent in the problem. Much less attention than one might expect is given to such fundamental notions as willingness to pay and willingness to accept compensation as measures of benefits.

Following a discussion of the concepts of marginal benefits and marginal costs in the context of climate change, the authors present some of the empirical estimates that have been made. They turn then to a discussion of some alternatives to CBA when full information is not available. These include the "precautionary approach," the so-called "safe minimum standard approach," and something called the "absolute safe standard approach." They then plunge into a discussion of the types of benefits that might make up something they call "total economic value." This includes direct benefits such as those to health, indirect benefits such as flood control, option values for future direct and indirect benefits, and finally, nonuse values such as the knowledge that a habitat might be preserved even though it would be of no direct or indirect economic value to the beneficiary. Finally, a fair amount of space is devoted to a discussion of the way in which multicriteria analysis might be used as an alternative to traditional CBA, illustrating this with a case study of the electric utility sector in Sri Lanka.

MINOR QUIBBLES

Briefly, and in the order in which they appear, it is useful to identify here assertions or characterizations in Chapter 5 that may be at variance with the way many other economists view the issues. First, the authors state that neither CBA nor the decision-analytic techniques that they (too blithely, I think) lump together with it can answer the question: Who should reduce emissions (IPCC 1996 ,149–50)? It would be much better to make three related statements. First, economic analysis can shed considerable light on both *where* emissions can be reduced most inexpensively and *which* adaptive strategies are the most inexpensive societywide. Second, there ought to be a rebuttable presumption that, if the decision is made to reduce greenhouse gas emissions and/or invest in adaptive measures, society ought to begin with the least expensive options first. Third and finally, this surely leaves open the question of who will pay the costs for these actions. The authors come around to this point—that is, to the distinction between where emissions ought to be controlled and who ought to pay for those controls—much later in their chapter, but long after creating what I believe is a misimpression about the utility of CBA (or cost-effectiveness analysis).

Also on page 150, the authors say, "Moreover, [CBA] presupposes that the relevant costs and benefits are those that ultimately affect human welfare," and leave it at that. This is confusing. Is this a criticism of the anthropocentricity of CBA? If so, the authors would have done well to pursue this point. Many noneconomists (and some economists, too, going back to John Stuart Mill) are troubled by the assumption underlying applied welfare economics that, if no human being cares enough about some type of environmental harm to pay something to reduce its likelihood, there is no benefit to doing so. It would have been useful to make this point explicit to decisionmakers without training in economics, and it is useful to remind economists what a strong assumption this is.

On the other hand, in making this statement, were the authors saying that the only things that count in a CBA are those things that affect humans directly? If so, the statement is manifestly false, since some people are willing to pay something to preserve wilderness areas they know they will never see. Since the authors discuss such "existence values" later in the chapter, I assume this is not what they meant on page 150, but it is unclear.

At the bottom of that same page, the authors note that "…CBA cannot provide answers about the optimum level of equity, in the same way it provides answers about the optimum level of economic efficiency." This is not at all surprising, I suppose, since *no* approach I know of can provide answers about "the optimum level of equity," whatever that means. Unless one assumes (or is provided by a higher authority with) a social-welfare function, it is impossible to say which of the infinite possible distributions of individual well-being is the "optimum." By the same token, it is not clear what the authors mean by the "optimum level of economic efficiency." If this is the allocation of resources that maximizes net benefits, better to say so and avoid the confusing jargon.

Later, on page 152, the authors say that "…sustainable development will require recognition of goals related to economic efficiency, social equity, and environmental protection." The authors are surely right here in pointing out that equity concerns must always be evaluated alongside those related to efficiency. But why do they separate environmental protection from economic efficiency in a chapter on CBA? What is the purpose of CBA, after all, if not to *incorporate* into efficiency calculations concerns about environmental protection and other social "goods" that will be insufficiently provided by unfettered markets on account of externalities, natural monopoly, or imperfect information? This bifurcation seems oddly out of place in a chapter on the role of CBA in climate decisionmaking.

When discussing irreversibility on page 154, the authors say that failure to reduce emissions now is troubling because "…once the effects of climate change become apparent, it will be too late to do anything about it." Not so. First of all, while carbon dioxide concentrations in the atmosphere do have considerable inertia, they also depreciate at a rate of about 1% per year, rather than remaining there forever. Second, there are ways to further reduce concentrations, for example through forestation. It is not helpful to suggest otherwise in a chapter likely to be read by policymakers.

In discussing local and regional environmental problems, also on page 154, the authors err in stating that these problems are easier to deal with than global change because "…the benefits of emissions reductions generally accrue to the same geographic areas as bear the costs." Since many air and water pollutants originate from large stationary sources—factories, power plants, and refineries, for instance—and since most of these very large sources produce things sold in national or even international markets, the costs of controlling many pollutants are distributed very widely, often much more broadly than the benefits.

In discussing so-called "below the line" control options on page 155 (that is, those which allegedly reduce carbon dioxide emissions and save money for those implementing the technological changes cited), the authors suggest that compact fluorescent lighting and vehicle inspection and maintenance programs (among several others) may be especially attractive. Perhaps, but both examples must be viewed warily. Compact fluorescent lighting has been criticized for the quality of the light it provides; if these criticisms are valid, consumers installing such lights would suffer at least some reductions in welfare that would have to be netted out of any cost savings that might result. Similarly, in evaluating vehicle inspection and maintenance programs, it is common to overlook the time and inconvenience costs that vehicle owners sometimes incur as part of these programs. Such welfare losses (as well as the costs of vehicle repair) must be racked up against the improvements in fuel economy that might result.

On page 162, the authors appear to commit what I regard as an unpardonable sin in CBA—using control costs as a proxy for benefits. (This sin is committed regularly by some state public utility commissions that are trying to assign values to the external costs—for example, environmental damages—associated with the electricity that would be generated by potential new sources.) Unless one is prepared to argue (as I am sure the authors are not) that, elsewhere in the economy, all pollutants are being optimally controlled, then using control costs as a proxy for benefits makes no sense whatsoever. It is easy to think of cases where the cost of doing something—inscribing my memoirs on the head of a pin, say—would be phenomenally expensive, yet worth virtually nothing since no one would be willing to pay much to have such a thing. By the same token, there are as many cases where trivially inexpensive measures—vaccinations, perhaps—give rise to very large individual and social benefits. The authors nearly fall into the same trap two pages later, incidentally, in discussing replacement cost as a proxy for the value of protecting natural areas. Errors such as these do not belong in a document likely to be read by a great many people seeking advice on CBA and climate change.

Finally, on page 166, the authors present a discussion of issues relating to discounting that I found confusing, not to mention seemingly at odds with the quite sophisticated treatment of that same subject in the preceding chapter of the Working Group III report. For instance, the authors say, "The discount rate denotes the social opportunity cost of capital," and later, "...the social discount rate should be equal for all investments." This is at variance with the view that all effects, environmental and otherwise, ought to be converted to consumption equivalents and that these consumption equivalents should then be discounted at the appropriate discount rate on consumption, with the latter reflecting the pure rate of time preference, the elasticity of the marginal utility of consumption, and the rate of growth of per capita consumption. It is unclear to me why the authors raised this issue in the first place, since their discussion is not only inconsistent with that in Chapter 4, but also with the section on discounting in the Summary for Policymakers of the Working Group III report.

MORE SIGNIFICANT PROBLEMS

In addition to these difficulties in Chapter 5, I was troubled by two broader issues that I believe merit mention here. The first pertains to the unique challenge the authors of Chapter 5 believe global climate change poses to CBA, while the second concerns several criticisms of CBA the authors ought to have given greater emphasis to.

The "Mother of All Problems" View

There is a belief among some (perhaps many), including the authors, that global climate change poses truly unique problems for CBA in virtually every respect, and that this ought to make us humble when thinking about the implications of global climate change for applications of CBA. I have a different view. While, certainly, climate change heavily taxes CBA in some respects, they are fewer than the authors suggest. Indeed, virtually all of the complications that climate change poses for CBA arise in other applications, even confining attention to the environmental area. (For an excellent series of case studies on CBA in the environmental context, see Morgenstern 1997.)

I hasten to add that I do *not* conclude from this that it is a breeze to analyze the costs and benefits of measures to limit emissions of greenhouse gases into the atmosphere, or to evaluate the pros and cons of adaptive strategies. Rather, I suggest that the problems in applying CBA to other environmental questions have been too little appreciated, even in cases where CBA has played an important role in decisionmaking. Let me elaborate, using the reasons identified in Chapter 5 that the authors feel make climate change the mother of all environmental problems to which CBA might be applied.

First, the authors identify, correctly, the complex chain of events that links emissions of greenhouse gases to the valuation of costs and benefits. Yet, how is this different from emissions of volatile organic compounds (VOCs) and oxides of nitrogen (NO_x) combining in the atmosphere in the presence of sunlight to produce ground-level ozone (smog), which in turn adversely affects human health and the environment? What distinguishes the chain of events in global warming from that associated with discharges of heavy metals into water bodies and their eventual effects on individual welfare? Nothing, so far as I can see. In fact, virtually every good textbook on environmental economics identifies this same sequence, identified in Chapter 5, as the chain of events one must consider in thinking about virtually *every* environmental problem. (For instance, see Freeman [1993, 31] for a description of this same chain of causality in his discussion of the use of CBA for *all* environmental problems.)

Next the authors point to the fact that carbon dioxide is a stock rather than a flow pollutant. But there is nothing unique about this. So, too, are PCBs, mercury and other heavy metals, organochloride pesticides, and other pollutants. In fact, one reason for the current concern about certain chemicals in the environment (that may disrupt the endocrine and other hormonal systems in the human body) is that these chemicals are alleged to be bio- and enviroaccumulative. All these substances pose different (and in some respects more difficult) problems for CBA, problems also posed by carbon dioxide and other greenhouse gases. The same is also true of the "inertia and irreversibility" the authors claim is unique to global climate change. Once generated, for example, radioactive wastes are "irreversibly" toxic for any meaningful period of time.

The authors quite correctly point out that greenhouse warming is a global problem and thus a more difficult problem to *solve*. Its global nature, however, does not necessarily make it a more difficult problem to which to apply CBA (although it does force one to struggle with the distribution of both costs and benefits across nations). In fact, the CBA done by the U.S. Environmental Protection Agency in support of its regulations phasing out the use of chlorofluorocarbons (CFCs), another truly global environmental problem, has recently been recognized as among the best ever performed by the EPA (Morgenstern 1997). The authors correctly point out that the benefits of any measures that might be taken to slow the accumulation of greenhouse gases in the atmosphere will be nonuniformly distributed. In fact, poorer and low-lying countries will enjoy disproportionately large benefits (since they will probably suffer the most if mitigating measures are not taken). Again, however, this differs not much from the prevention of stratospheric ozone depletion, although the former is a harder problem than the latter because some nations would benefit from a warmer climate, while no nations could be presumed to benefit from increased exposures to damaging ultraviolet radiation.

The authors of Chapter 5 also assert that climate change is a difficult case for CBA because, as they put it, "...actual impact data are scarce," by which they appear to mean that we have no dead bodies (or ecosystems) as evidence of the seriousness of the problem. Yet I see no difference between this case and, say, the case for controlling ground-level ozone on account of concerns about its adverse effects on human health and the environment. True, some epidemiological evidence links exposures to ozone to acute respiratory disease, and animal toxicological evidence suggests a link between ozone and chronic illness. But the epidemiological evidence is limited by the frequent lack of controls for other possible explanatory factors, and the animal evidence is tainted by the extreme difficulty of extrapolating from test animals to humans.

The situation is no different for many other environmental problems. Only rarely, in fact, do we have very convincing evidence that if a regulatory action is taken, a difference is sure to be observed in a problem of concern. The fall in blood lead levels in children following the phased removal of lead as an additive in gasoline stands out, not only because of its obvious health significance, but also because of the relatively rare cause-and-effect relationship it demonstrated. When other regulatory action has been taken in the past, and even when it was followed by improvements in human health and/or the environment, it has generally been very difficult to partition the effects of policy changes from those of other factors that have stubbornly refused to "stay constant."

Nonlinearity is another unique feature of global climate change, the authors assert. Tell that to those modeling the effects of reductions in NO_x emissions on the formation of ground-level ozone. In this latter case, NO_x emissions in the presence of VOCs increase ozone concentrations under certain conditions, but reduce ozone under other, often not too different conditions. Nonlinearities characterize the relationships between discharges of biochemical oxygen demand and water quality, between exposures to certain carcinogens and the likelihood of carcinogenesis, and between emissions of CFCs and stratospheric ozone levels. The nonlinearities believed to characterize climate change make it a more difficult problem, to be sure, but they do not make it a unique problem. There are clear parallels to other problems to which CBA has fruitfully been applied.

Finally, throughout Chapter 5 the authors repeatedly emphasize the uncertainties associated with climate change. These uncertainties pertain to the future level of emissions of greenhouse gases; the links between emissions and atmospheric concentrations, between concentrations and global average temperature change (as well as the regional changes in weather that might result), between these latter changes and such things as agricultural output, forest productivity, disease vectors, and so on; and, finally, the welfare implications of the physical changes triggered by a warmer atmosphere.

To be sure, these uncertainties are formidable. In fact, I am willing to concede that we *probably* know less about the key links in the chain of causality for this problem than we do about the analogous links for any other potentially serious problem. However, the problems associated with hormone-disrupting chemicals, bioengineered organisms, and habitat alteration all have uncertainties potentially as troubling as those associated with climate change. Nevertheless, the authors of Chapter 5 create the impression that the uncertainties associated with climate change are so formidable that CBA is of relatively little use, despite their occasional statement to the contrary. This leads them to a discussion of such alternatives as the precautionary approach, the affordable safe minimum standard, and the absolute standards approach (see 159–60).

The authors' discussion of these alternative approaches to decision-making is unlikely to be of much help to policymakers, I fear. I say this because there is precious little guidance in their discussion about how one might make operational progress in utilizing either the precautionary approach or the absolute standards approach. Consider first the precautionary approach. According to the authors' description, this involves identifying some level of control of greenhouse gases, but not so stringent a level that it would entail excessive control costs. Yet if one is thinking about the point at which the marginal cost of control curve begins to rise sharply in order to implement the affordable safe minimum standard approach, why not puzzle also over the location of the marginal benefit curve? Doing so would allow one at least to engage in speculation about the level of control that maximizes net benefits.

Finally, suppose one is enamored with what is called the absolute standards approach, in which some level of carbon dioxide in the atmosphere is identified as being too much. How does one decide what absolute standard is best? In the context of climate change, this would mean deciding that, say, no more than 500 ppm CO_2 in the atmosphere shall be tolerated (that would be the "absolute standard" to which the authors refer). In thinking about setting such a standard, and in the absence of firm evidence about threshold concentrations above which things "go to hell in a handbasket," would one not think about the costs associated with alternative absolute standards and the incremental protection that successively more stringent standards would provide? In other words, would one not engage in the same kind of thinking that would characterize CBA? It is hard to answer this question other than in the affirmative, I would think. Matters are different, of course, if a physical threshold can be identified, though even in that case it would seem to me decisionmakers would still want to think about the nature of the damage associated with that threshold and the costs of avoiding it.

To summarize this discussion, then, I do not take issue with any of the issues the authors use to characterize the problem of climate change. I do however, believe they create an exaggerated picture of the uniqueness of these problems. Rather, it would seem, virtually every issue they raise can be seen in other environmental areas, and some problems—stratospheric ozone depletion, long-range transport of acidifying air pollutants, and even ground-level ozone, for instance—display most or all of the same troubling aspects. To be sure, however, the intergenerational issues associated with the climate-change problem—and the discounting questions they raise—are particularly knotty, as is the uneven distribution across nations of the potential costs and benefits of mitigation actions.

Criticisms Underplayed

Given what I take to be the authors' at least mild antipathy toward CBA, I was surprised at their virtual silence on several issues on which I think CBA is vulnerable. Consider, first, the estimation of costs. At several points throughout their chapter, the authors raise difficulties with cost estimation, but these have more to do with engineering problems than conceptual underpinnings. That is, the authors (correctly) cite the debate between the "top-down" modelers (whose estimates tend to be on the high side) and the "bottom-up" estimators (who, because of technological optimism, tend to regard the likely eventual costs of CO_2 mitigation as quite affordable) as evidence that costs are difficult to estimate accurately.

Yet they miss altogether what I take to be a more fundamental difficulty with cost estimation for CBA. This is the fact that, conceptually speaking, costs ought to be measured in the same way that benefits are in CBA. Thus, the same difficulties that plague benefit estimation also make it more difficult than many suppose it to be to estimate costs. To elaborate, the correct measure of costs for CBA is the aggregate compensation that would have to be paid after enacting a policy change to make everyone harmed by that change as well off as they were before its implementation. This includes those who face higher prices as a result of a policy change (limitations on CO_2 emissions, say), those who have suffered job losses or reduced incomes, those who no longer have access to products they consumed prior to the change, and so on.

To estimate these costs, one has first to model the mechanism through which the initial incidence of control expenditures is transmitted through the economy to other sectors. Those other sectors would include some industries that bear none of the initial costs (higher costs to the financial services sector, for instance, as a result of controls imposed on the electricity-generation sector that supplies inputs to financial services). Ideally this modeling should be accomplished with a general equilibrium model such as those used by Hazilla and Kopp (1990) or Jorgenson and Wilcoxen (1990).

Next, costs should be calculated using consumer expenditure functions, so that compensation required can be measured in a manner consistent in a welfare-theoretical sense with that used for determining benefits. This is no easy matter, requiring again the kind of models used by the researchers cited above. It will be made much more difficult in the context of global climate change because of the paucity of data on the preferences of those in developing countries.

A second, even more difficult problem with the application of CBA to climate change is this: estimating the damages associated with unchecked global warming. Another way to express this problem is to say that it will be very difficult to measure individuals' willingness to pay (WTP) for the favorable impacts of mitigation measures or adaptive policies. This is because individuals' willingness to pay depends critically on the distribution of income, so that the more money any given individual has, the more able he or she is to pay for a given environmental improvement. This dependence is the Achilles' heel of CBA when viewed by noneconomists and makes CBA controversial even when applied within a developed country. It is even more problematical when CBA is used to analyze problems that spill across international boundaries.

It is easiest, perhaps, to illustrate this point using WTP to reduce the risk of premature mortality. In the United States, cross-sectional analyses of compensating wage differentials across occupations (almost exclusively in manufacturing industries) suggest that workers are willing to accept an additional 0.001 in annual mortality risk for between $2,000 and $8,000 per year, with $4,000 being the median value (see Viscusi 1992). This implies a value for a "statistical life" of about $4 million, the value used by the Environmental Protection Agency in its applications of CBA to life-saving programs. And while there are respects in which this value is somewhat unsatisfactory (it is based on premiums for risks that are borne more voluntarily than most environmental risks, for instance), there is surprisingly little controversy within government about its use in modern applications of CBA.

Even for U.S. applications, however, it seems to me that there ought to be more scrutiny of this and other values used in CBA, given their dependence on the existing distribution of income. For instance, imagine a pollutant that affects those in both a very wealthy and a very poor neighborhood. Because those in the poor neighborhood have very low incomes, their WTP to reduce the concentration of the pollutant will in general be rather low (they simply cannot afford to pay more than $50 per year, say, to reduce their annual mortality risk from the pollutant by 0.001). Thus, the value of a statistical life saved in the poor neighborhood would be $50,000. Suppose those in the wealthy neighborhood—Aspen or Beverly Hills, perhaps—would be willing to pay $15,000 for the same reduction in risk; the value of a life saved here would be $15 million—

three hundred times as much as in the poor area. It follows, then, that—distributional considerations aside—society would invest resources more efficiently saving 1 life in swanky Aspen or Beverly Hills than 299 lives in a poor neighborhood.

There are reasons for defending CBA in spite of such implications, reasons every good economist ought to know. In the case of this objection to CBA, the counter argument is this: if we do not use individuals' own valuations of reductions in the risks they face, where will we derive these values? But this aspect of CBA is much more troubling, to me at least, than virtually anything the authors of Chapter 5 raised in their discussion. If the purpose of that chapter was to provide policymakers and others with a sense of the strengths and weaknesses of CBA in climate policy analysis, they should have devoted much more time to this issue.

This point is strengthened considerably given the international spread of the likely costs and benefits of mitigation measures. How do we evaluate policies that may prolong the lives of U.S. citizens, say with an assumed per capita value of $4 million, with those that may save lives in Bangladesh or Sri Lanka? In those latter countries, a statistical life could be valued in the hundreds of dollars based on traditional WTP measures. Indeed, one persistent criticism of early drafts of the IPCC report was lodged by a group concerned about this very problem. Are we willing to live with the consequences of a CBA driven by individual preferences regarding life-saving, ecosystem protection, and so forth? (This point is discussed at length in Chapter 6 of IPCC 1996.)

To illustrate how difficult it is to escape this problem, suppose we were to select an "average" value for a statistical life that puts the worth of lives saved in developing countries at closer to the value(s) that would be appropriate in developed countries. This creates another kind of problem for the application of CBA to climate-mitigation policies. If we assign a developed-country value to lives saved in developing countries, there will be many policies unrelated to climate that will save lives immediately (improved water supply, wider spread vaccinations, and so on). These might be so attractive that developing countries would concentrate all their attention on programs that save lives in the short term, and pay little or no attention to climate change or other problems where the life-saving benefits will not be felt for many years.

This is the kind of problem, it seems to me, that was ripe for discussion in Chapter 5. It is important but rather subtle, and it is of great relevance for environmental and other problems that spill across national boundaries. Such a discussion might have included a defense of income-based WTP (saying, for example, we don't know what people would be willing to pay under other, hypothetical distributions of income, for instance) and might also have raised questions about how we might value the beneficial effects of climate change without resorting to the preferences of those who would be affected.

A Greenhouse Referendum?

In one final respect I think Chapter 5 might have made a more significant contribution to the debate over CBA and its role in evaluating climate-change policies. It concerns the way that cost-benefit deliberations are envisioned in Chapter 5, and how they might be more productively viewed. (To be fair to the authors, however, it involves a very different approach to climate decisionmaking. For that reason, they cannot be blamed for not having investigated it.)

To illustrate, the authors of Chapter 5 buy into what might be called the damage-function approach to CBA and climate change. That is, they describe a CBA approach in which one first ascertains the time path of costs and the physical damages that would be prevented by measures to slow emissions of greenhouse gases, then assigns dollar values to these costs and benefits (the latter being the damages avoided), and finally discounts the stream of future costs and benefits back to the present using an appropriate discount rate. There is nothing wrong with this approach at all. In fact, it is the most common way to formulate an application of CBA, particularly in the environmental field.

But it is not the only way to structure the problem, and, I would argue, it may not be the most useful way to view it. Rather, why not view the climate-change problem as involving a decision about "social insurance"? Such an approach would be predicated on the view that we will never be able to develop very precise estimates of the future benefits associated with preventing climate change. Moreover, according to this view, even if we could develop monetary estimates of future benefits, the problem of selecting a discount rate would make traditional CBA intractable.

The question to be answered if we accept this view is the following. How much are members of the present generation willing to pay to reduce the likelihood of a stream of adverse effects (and some positive effects) happening in the future to an entirely different group of people, most of whom are not now alive and, when they are, will be living in different countries? If aggregate WTP summed over all the citizens of the world is greater than the cost of the corresponding risk reduction, then it makes sense on efficiency grounds to enact the insurance policy. If aggregate WTP is less than the cost of the insurance policy, then the policy would be socially inefficient and should not be enacted unless it were sufficiently attractive on other grounds.

There appear to be several attractive features to viewing the CBA decision in this way. First, it mimics the political decision that must be made if we are going to act to slow the accumulation of greenhouse gases in the atmosphere. That is, even if members of future generations will value extraordinarily highly the beneficial effects of mitigation measures taken today, these measures will not be enacted until and unless members of the present generation decide to bear the costs of initiating such measures. The question each respondent will ask himself is, "Will the good I think this policy will do (mostly for others at a later point in time) outweigh the cost to me of putting the policy in place?"

Second, structuring the problem in this way alleviates the need to guess what values future generations will place on the costs and benefits they will bear, as well as the need to select a discount rate to bring the stream of future costs and benefits back to the present. Rather, it requires describing as completely as possible the likely stream of future physical benefits, ranging from the near term into the distant future, and then attempting to elicit accurately each individual's WTP to have this stream. Note that this allows each individual to attach his or her own discount rate to the stream of future effects, thus alleviating the need to choose a single (and therefore artificial) social discount rate. Individuals with very low discount rates will, other things being equal, have higher valuations of the benefit stream. People with high discount rates, on the other hand, will attach relatively lower values.

I hasten to add—though it is probably not necessary—that this approach is no panacea. It requires, after all, describing accurately to members of the present generation the likely future effects of climate mitigation or adaptation policies (as well as any side–costs and benefits that might occur immediately), so that they will know what they are buying. More importantly, it requires ascertaining their WTP for this stream of future "goodies."

The uncertainties surrounding future damages from climate change—of which the authors of Chapter 5 make much—would make it *very* difficult to present the nature of this stream of avoided damages with much accuracy at all. What is needed, of course, is a very thorough description of the expected result of a policy change. While it would have to be much, much more detailed than this, such a description might start in the following way:

"Some scientists believe that climate-mitigation measures will forestall or prevent altogether such otherwise likely catastrophes as the inundation of small island nations, epidemics that might take many thousands or even millions of lives, and the eradication of countless species. Other scientists believe such drastic events as these are unlikely to accompany global warming but point to other quite serious effects, including the disruption of global agriculture and an increase in the severity and frequency of extreme weather events such as hurricanes, droughts, and floods."

Clearly, more comprehensive information would have to be provided about the many other potentially favorable effects of greenhouse gas mitigation measures, but this gives a sense of the way one might describe the beneficial attributes of a mitigation policy.

The problem does not stop with providing this description, however. The more difficult problem, of course, is determining people's WTP for an action or series of actions that would reduce the likelihood of the adverse events. Unfortunately, observable market behavior contains few traces of the value(s) that individuals place on the well-being of future generations. The difficulty of observing WTP directly raises the issue of contingent valuation (CV). This is not the place to raise the problems and promise of CV (see Portney 1994, Diamond and Hausman 1994, and Hanemann 1994 for a discussion of these issues.). Suffice it to say that, *if* there was agreement on a set of conditions under which CV was capable of eliciting useful information about WTP, this would be the paradigmatic problem for which it would be used.

Despite its difficulties, I see this "insurance policy" view of the climate-change problem as a useful alternative to the damage-function approach the authors have assumed. The former still lies within the CBA paradigm: it poses the problem of ascertaining how much people will pay for a future stream of benefits and comparing that amount (however divined) to the cost of the insurance policy.

Because the authors of Chapter 5 frequently mention risk aversion in their discussion of CBA and climate change, it surprises me that they neglected to mention a conception of the problem as akin to an insurance purchase. Admittedly, one would face very serious difficulties in attempting to implement the climate-referendum approach I suggest here. Nonetheless, I think it has much to recommend it—not the least of which is that it represents the fundamental question that must be answered if climate-change policies are to be adopted.

CONCLUSION

Having spent the better part of this chapter identifying respects in which I think the authors of Chapter 5 might have improved their chapter, let me return now to the point I made in opening this paper. Describing the strengths and weaknesses of CBA as applied to global climate change is a difficult assignment under the best of conditions. To have to do so in thirty to forty pages invites the kind of second-guessing I have engaged in here. In many respects, the authors hit the right high spots and conveyed a sense of the difficulty in using CBA for climate change. Had our roles been reversed, they could have fairly faulted me for being less than comprehensive, and for errors of omission and commission, as well, no doubt.

My preference would have been for a chapter that was at once both more open about the glaring weaknesses of CBA (to outsiders, at least), but also stronger in its insistence that CBA has an important role to play in evaluating policy options in the climate-change debate. I hope that both points are made forcefully by those involved in drafting future reports on the uses of economics—and especially CBA—in analyzing climate-change policies.

REFERENCES

Diamond, Peter A. and Jerry A. Hausman. 1994. Contingent Valuation: Is Some Number Better Than No Number? *Journal of Economic Perspectives* 8(Fall): 45–64.

Freeman, A. M. 1993. *The Measure of Environmental and Resource Values.* Washington, D.C.: Resources for the Future.

Hanemann, Michael. 1994. Valuing the Environment Through Contingent Valuation. *Journal of Economic Perspectives* 8(Fall): 19–43.

Hazilla, Michael and Raymond J. Kopp. 1990. Social Cost of Environmental Quality Regulation: A General Equilibrium Approach. *Journal of Political Economy* 98(August): 853–73.

IPCC (Intergovernmental Panel on Climate Change). 1996. *Climate Change 1995. Economic and Social Dimensions of Climate Change: The Contribution of Working Group III to the Second Assessment Report of the Intergovernmental Panel on Climate Change.* Edited by J. P. Bruce, H. Lee, and E. F. Haites. Cambridge: Cambridge University Press.

Jorgenson, Dale and Peter Wilcoxen. 1990. Environmental Regulation and U.S. Economic Growth. *RAND Journal of Economics* 21(Summer): 314–40.

Morgenstern, Richard D. 1997. *Economic Analysis at EPA: Assessing Regulatory Impact.* Washington, D.C.: Resources for the Future.

Portney, Paul R. 1994. The Contingent Valuation Debate: Why Economists Should Care. *Journal of Economic Perspectives* 8(Fall): 3–17.

Viscusi, W. Kip. 1992. *Fatal Tradeoffs.* New York: Oxford University Press.

Comments

Applicability of Cost-Benefit Analysis to Climate Change

Ferenc L. Toth

It is no mean task to add something sensible to the discussion about an issue already addressed by the carefully selected team of experts who contributed to Chapter 5 of the IPCC Working Group III report (IPCC 1996) and in this volume by Paul Portney, whose numerous contributions to the subject of cost-benefit analysis (CBA) are also well known and highly valued. Yet the subject is enormous and highly controversial. Consequently, the debate will certainly continue for years to come. In this commentary, I would like to offer a few points reflecting both on Chapter 5 and on Portney's critical review.

It is appropriate to start by saying that I share Portney's conviction regarding the difficulty of the task that the authors of Chapter 5 faced. The reasons are not exclusively the complexity of the subject and the limited space available for addressing it. Greenhouse-gas (GHG) emissions and resulting changes in the climate system have engendered apocalyptic visions of possible impacts on nature and society. Any economist would pause for a second before proposing the use of off-the-shelf techniques of CBA for guiding social decisions under these circumstances. The authors of Chapter 5, of course, know the subject well and recognize the usefulness of CBA, but their formulation is often far too defensive.

This posture is not necessarily due to what Portney calls "the authors' at least mild antipathy toward CBA." It might simply reflect the substan-

FERENC L. TOTH is project leader at the Potsdam Institute for Climate Impact Research (PIK) in Potsdam, Germany.

tial body of literature that rejects the relevance of cost-benefit principles for the climate-change problem offhand. Chapter 5 thus has a split mentality: its authors are trying to argue for the usefulness of CBA, but they go overboard, in my view, in trying to preempt criticism by diluting the strengths of traditional CBA and incorporating other techniques under the heading of "modern CBA." My suspicion is that many of the imprecisions and slips noted by Portney are due to this basic attitude.

The key question is to what extent results from traditional, project-level CBA should guide climate policy. Should one take seriously the proposal to rank climate policy on the list of other environmental issues strictly on the basis of a single CB indicator? Or should one acknowledge distinguishing characteristics of the climate issue and consider factors beyond the comfortable and well-understood scope of economics? Further, are there meaningful ways to combine various perspectives?

Chapter 5 raises four "central questions" about managing the global climate-change problem and evaluates the applicability of CBA to answering them. The questions relate to GHG emissions reduction: how much, when, how, and by whom? The authors then specify seven "unique features" of the climate problem to demonstrate the need for extending the arsenal of traditional CBA. Three such "extensions" are defined as "techniques of modern CBA," and it is concluded that they are "the best framework for identifying the essential questions" of climate policy.

The central thesis of Chapter 5 is that "traditional project-level CBA is too narrow to be relevant for evaluating climate change issues" (IPCC 1996, 152). To support this claim, the authors provide a list of characteristics which they claim are unique to the climate problem. Portney's review lists a broad range of cases in environmental management in which not only single "unique features" but various subsets of their combinations were present—yet traditional CBA provided useful information for the policy process. The most obvious counter example, stratospheric ozone depletion, is characterized to varying degrees by the same features. A short appraisal of the *ex ante* and *ex post* (that is, pre- versus post-Montreal Protocol) CBAs of CFC control and their repercussions in policy processes would have been very useful.

Admittedly, the listed features of the climate-change issue make application of traditional CBA difficult. In my view, however, the authors' response to this difficulty is more confusing than helpful. They claim to extend the concept of CBA to allow for addressing problems that traditional techniques have difficulties with. Their first extension is, in reality, a contraction. Recognizing the complications involved in estimating damages from climate change and assigning monetary values to climate-change impacts outside the national accounts as well as responding to the requirements of the policy process, several studies estimate the costs associated with various emissions and/or concentration targets. Many of these studies also involve optimization processes with respect to achieving those predefined targets most effectively. This amounts to the easier half of a CBA in the climate case and thus can hardly be considered an extended CBA.

Chapter 5 proposes *multicriteria analysis* (MCA) as an alternative response to the difficult and controversial issue of economic valuation of many climate impacts. While MCA clearly helps avoid the pricing problem, assigning relative weights to various attributes may turn out to be equally controversial. Nevertheless, Chapter 5 correctly points out the usefulness of creating trade-off curves. The sensitivity of these curves to various ways of selecting and defining attributes and their different weighting might provide valuable insights that usefully complement results of CBAs. The example presented in Chapter 5 is an unfortunate one. It includes CO_2-mitigation options in Sri Lanka's electricity production (an option on the cost side of the climate problem). In this case, MCA is not an "extension" of CBA either. The latter attempts to consider costs of impact and costs of control simultaneously.

The next step presented in the WGIII chapter for extending traditional CBA is motivated by the various types of uncertainties plaguing the climate-change problem. The authors advocate *decision analysis* as the proper tool for incorporating those uncertainties, especially the cost of irreversible effects, into the CBA. While the discussion of uncertainty contains a lot of useful material, it fails to emphasize a major analytical challenge of applying CBAs (whether traditional or broadly defined). This challenge relates to abrupt climate change or discontinuous impacts of a smooth climate change in major geophysical systems, such as the thermohaline circulation, the South Asian monsoon system, the El Niño southern oscillation, and the like. These events cannot be represented by a simple uncertainty band around the damage (benefit) function. The chapter points out that these low-probability, high-impact events often generate counterintuitive public reaction and result in extremely risk-averse behavior. Neither customary sensitivity analysis nor traditional decision analysis offers much help in addressing these issues. It might have been valuable if the authors had provided a literature review of the topic or had pointed out the lack of literature on the subject.

This, then, completes the list of techniques of "modern" or "broadly defined" CBA. The obvious immediate reaction is that, if other fields of application (that is, most social-decision problems) are characterized by similar difficulties and traditional CBA has provided useful information for decisionmaking, why assume that its careful application to the climate case would necessarily produce irrelevant results? In fact, early applications by Nordhaus (1991) and Cline (1992) were very helpful in organizing available information and in guiding later research in the field of *integrated assessment modeling*, which seems to have become a cottage industry and which is described by Charles Kolstad in Chapter 9 of this volume. Nevertheless, this is not to imply that CBA results should be the sole basis for making social decisions on climate change.

At a more general level, one should remark that Chapter 5 cites no other publications in which these various techniques (cost-effectiveness analysis, multicriteria analysis, decision analysis) are simply considered to be modern CBA techniques. As noted above, in most applications of what the chapter calls traditional, project-level CBA, analysts face similar difficulties. Attempts have been made (and documented) to overcome the perceived shortcomings even in those contexts. Readers of the IPCC report who are less familiar with the economics literature of the climate problem and with the underlying conceptual and methodological debates may find this extension of CBA more confusing than helpful.

Paul Portney's second major comment relates to another deficiency in Chapter 5, namely its failure to address the conceptual difficulties involved in estimating both costs and benefits using the correct measures. The basic question in CBA is how a particular project will affect the welfare of each individual concerned, that is, what amount would these individuals be willing to pay to obtain the benefits or to avoid the costs. Establishing these values in itself presents enormous difficulties in CBA. Deducing the change in social welfare from changes in the welfare of all individuals poses problems of similar magnitude. Unless income is optimally distributed (meaning that each person's dollar is equally valuable), we need to value the poor individual's extra dollar higher than the rich individual's.

The example of deriving a value for a "statistical life," presented by Portney, makes this difficulty very clear. Willingness (ability) to pay for mitigation and adaptation depends critically on income distribution. This issue becomes crucial when health-related impacts and losses of life are evaluated. Reactions to other chapters in the IPCC Working Group III report have shown that these are the most contentious issues in valuing climate-change impacts. Not surprisingly, the major source of differences in damage estimates is also associated with health and mortality impacts.

Willingness to pay and valuation issues are characterized by a double twist in the climate context. Income differences across countries and related international equity issues are highlighted in both impact studies and emissions-reduction analyses. In contrast, these issues have attracted much less attention in studies on the distribution of burdens of climate change or GHG-abatement policies among income groups within individual countries. Portney's proposal to consider a different conception of the CBA problem in the context of climate change is certainly appealing. Yet, in addition to the problem of providing an accurate presentation of likely future effects of adaptation and mitigation policies, determining the present generation's willingness to pay would be troubled by difficulties similar to those plaguing the damage-function approach to CBA, except that part of the problem would be shifted to the domain of contingent valuation.

To return to the initial questions: Are techniques of CBA applicable to the climate-change problem at all? If so, to what extent should their results guide policy? Three major lines of thought have emerged in the debate over the past few years.

The first maintains that, because of the very nature of these long-term issues, the impacts (that is, benefits) will come decades later. This delayed effect leaves ample time to revisit the issue regularly in the future. The implication is that traditional, off-the-shelf CBA is appropriate for policy analysis even in this case. The discounting technique and the discount rate should therefore be the same for climate change as for any other public policy issue. Regular reassessment of the issue will ensure that policymakers recognize any problems in due course and revise policies accordingly.

The second line of thought acknowledges that CBA is appropriate for addressing climate policy but makes a more concerted effort to bring distant economic losses due to global warming to the attention of present-day decisionmakers. The approach proposes to use lower discount rates to value faraway future impacts. Increasing evidence (for example, Nordhaus 1996), however, shows that fudging the discount rate does not help either to save ecological treasures in the distant future or to achieve efficient abatement policy.

The third group maintains that if there are hard-to-value assets or highly valued environmental components at risk and/or the inertia of the underlying biogeophysical system is such that there is a severe danger of going beyond a point-of-no-return, then the cost-benefit argument has only limited validity. The best and most economically efficient strategy in this case is to define long-term environmental goals and work out the optimal cost-effective policy for reaching them.

I have been arguing along the lines of this third approach. CBA is nevertheless an important source of information. Keeping in mind all their drawbacks and deficiencies, cost-benefit ratios for climate change (both the damage-function and the WTP kind) are useful for comparison with cost-benefit indices derived for other environmental issues and social-policy problems. Cost-benefit ratios, however, should not be the sole basis for social decisions. Analysts have the responsibility to help policymakers and other social actors define their long-term environmental targets. With the current state of our knowledge about climate-change impacts ranging from profound uncertainties to outright ignorance, providing the necessary information to set those environmental targets is extremely difficult at best and completely impossible according to many. Nevertheless, systematic attempts to search for the "ultimate reasons" for climate protection in various impact sectors are useful in sorting out thorny issues about climate vulnerability, impacts, adaptation, and the assessments thereof.

This is the very strategy the Potsdam Institute for Climate Impact Research is implementing in its project on Integrated Assessment of Climate Protection Strategies (ICLIPS). The approach is a bi-directional analysis—from tolerable climate impacts to costs associated with emissions-reduction measures that keep the climate system within the derived climate window, and vice versa. The project seeks to define climate-response functions for various climate-sensitive sectors. Social actors can then use the response functions to define their perceived tolerable levels of climate impacts. These constraints would then define regionally tolerable climate windows. By using an appropriately formulated integrated climate and economic model, cost-effective emissions paths will be derived that keep the global climate system within those tolerable windows. In the opposite direction, the model should be able to compute through the traditional analytical path from emissions scenarios to climate change to damages in numerous impact sectors. Costs associated with various tolerable climate windows as well as the benefits secured by them in terms of natural biophysical units could then be compared in a further analysis. In working out the cost-effective emissions path, of course, costs in various future points of time would need to be compared. This intertemporal optimization problem would adopt discount rates consistent with both economic theory and empirical observations.

It is clear from both Chapter 5 and Portney's critical review that, despite significant improvements in, and a huge body of practical experience with, CBA over the past two decades, our tools are still imperfect. But so are others presented as extensions of traditional CBA. Further improvements are badly needed. For example, cross-fertilization by adopting specific concepts and techniques across the various approaches might be useful. Lumping them together, however, is confusing at best. In some cases, the various approaches are based on different paradigms or partially contradicting theoretical underpinnings.

Despite its shortcomings, Chapter 5 contains a significant body of useful material about framing the climate problem for economic analysis. Yet the title is misleading. The chapter conveys more information about policy-analysis techniques appropriate for addressing climate-change issues than it does about the applicability of CBA-proper as most economists and policy analysts know it. Portney's review, albeit selective, is helpful in putting a few key issues addressed by Chapter 5 in perspective as well as in providing interesting ideas for future research. It would not be surprising if the next IPCC assessment included reviews of studies that implemented and elaborated on these ideas.

REFERENCES

Cline, W. R. 1992. *The Economics of Global Warming*. Washington, D.C.: Institute for International Economics.

IPCC (Intergovernmental Panel on Climate Change). 1996. *Climate Change 1995. Economic and Social Dimensions of Climate Change: The Contribution of Working Group III to the Second Assessment Report of the Intergovernmental Panel on Climate Change*. Edited by J. P. Bruce, H. Lee, and E. F. Haites. Cambridge: Cambridge University Press.

Nordhaus, W. D. 1991. To Slow or Not to Slow: The Economics of the Greenhouse Effect. *Economic Journal* 101(6): 920–37.

———. 1996. Discounting and Public Policies that Affect the Distant Future. Paper presented at the EMF-RFF Workshop on Discounting, November, Washington, D.C. In *Discounting and Intergenerational Equity* (forthcoming), edited by John R. Weyant and Paul R. Portney. Washington, D.C.: Resources for the Future.

5

Greenhouse Policy Architectures and Institutions

Richard Schmalensee

The IPCC Working Group III report (IPCC 1996, hereafter "the report") is a remarkable piece of work. It is little short of amazing that such a large group of authors and reviewers, operating under such stringent procedural and substantive constraints, could produce such a high-quality document. A policymaker seeking guidance on the most useful things to do over the next few years could learn much from this report about the economic dimensions of climate change and about the design of efficient environmental policies in general. In aggregate, the report does a very good job of providing a comprehensible overview of a wide expanse of relevant intellectual territory, some of it no doubt politically treacherous. The exposition is usually clear and sometimes even elegant, and, while the report is not fully internally consistent, the level of consistency attained is remarkable in light of the production process involved.

This said, however, I believe that a policymaker seeking guidance for near-term actions would likely come away from the report disappointed—or, in the worse case, misled—on some important issues. The main problem is errors of omission, not errors of commission. The report presents a great deal of information that would be useful to a climate czar making a once-and-for-all global policy choice, but there is no such czar, and the key near-term choices involve institutional designs and policy architectures that will help frame future policy debates. Moreover, the report pays insufficient attention to the long-term consequences of possible near-term choices and fails to develop key analytical points of which policymakers should be aware.

RICHARD SCHMALENSEE is Gordon Y Billard Professor of Economics and Management, Massachusetts Institute of Technology.

These omissions reflect in part the authors' main assignment, assessment of the available literature. They also reflect the report's linkages with ongoing international negotiations, which have been fixated on setting near-term emissions limits for a relatively small number of industrial nations (the so-called Annex I countries). As will become clear in what follows, I view this fixation as an unwise architectural decision.

The report concentrates on questions regarding policy instruments and their optimal levels—questions which are central to the literature on environmental economics and to the domestic policy debates that the literature mainly seeks to inform. Unfortunately, the international negotiations have been conducted as if the same questions were also central there. As I argue in what follows, however, the arena in which international climate-change policy is shaped differs fundamentally from those in which domestic environmental policies are determined, particularly at this early stage in the process. Moreover, the climate issue differs in important and perhaps fundamental ways from issues that have been addressed (with mixed success) by other environmental treaties. The authors of the Working Group III report, and particularly the authors of Chapter 11, "Policy Instruments," have naturally written more about what we know than about what we need to learn. Unfortunately, the latter is presently more important. By describing important gaps in our knowledge clearly, the report could have made a significant contribution to the intelligent determination of research priorities.

The assertion that the report pays inadequate attention to issues related to the near-term policy agenda plainly rests on a particular view of that agenda. The next section outlines that view.

THE CLIMATE ISSUE TODAY

There appears to be near-universal agreement regarding several key features of the climate-change issue, most of which are developed by the report.[1] First, the relevant economic and physical processes operate globally and over decades rather than years. Most plausible emissions scenarios involve a significant human-induced increase in radiative forcing over the next century, with much of the increase coming from emissions of countries that are not now wealthy.[2] Today's emissions will affect the chemistry of the global atmosphere for a century or more and appear likely to affect climate for even longer. Today's investments in research and in energy-sector capital will shape economic activities and affect emissions for at least several decades. Because these lags are very long, a range of current actions have climate-related consequences that can usefully (though only approximately) be treated as irreversible. Few observers foresee substantial climate change for at least several decades, after emissions and atmospheric concentrations have increased substantially.

Second, important and probably long-lived uncertainties are ubiquitous. There are important unanswered questions involving the atmospheric chemistry of trace gases and aerosols, fundamental climatic processes, future emissions, future technologies, the costs of abating emissions, and the costs of adapting to climate changes. Despite the language of the Framework Convention on Climate Change, which calls for "stabilization of greenhouse gas concentrations in the atmosphere at a level that would prevent dangerous anthropogenic interference with the climate system" there are no known thresholds in that system. On the other hand, there are reasons to think that some (unknown) level of climate change could lead to a rapid and discontinuous change in the pattern of ocean currents. Though emissions of many gases and aerosols apparently affect radiative forcing, significant uncertainties attach to the sources and effects of some of these emissions.

Third, the climate issue involves potentially huge stakes. On the one hand, the very survival of the human race depends on the earth's climate, so that experimenting with that climate seems insane. On the other hand, stabilizing atmospheric concentrations of greenhouse gases within a century or so, even at levels well above today's, is likely to be very expensive. It will almost certainly require reducing global emissions of carbon dioxide (CO_2) well below current levels.[3] At the very least, since most anthropogenic CO_2 emissions are produced by combustion of fossil fuels, reducing global emissions would likely prevent today's poor nations from becoming wealthy using currently known technologies. Reducing global CO_2 emissions substantially, relative to trend, would require transforming the energy systems of both developed and developing nations and, as Chapter 9 of the report indicates, would likely involve annual costs on the order of several percent of world income. Such costs would dwarf those of eliminating CFCs from the global economy. The total direct cost of all current U.S. environmental programs, many of which are extremely controversial, comes to only about 2% of GDP. Agreeing to incur incremental costs of this magnitude without clear evidence that *any* benefits will result also seems a bit insane—particularly from the viewpoint of poor nations with serious and more immediate environmental problems.

Fourth, analyses of globally optimal climate policies generally do not support imposing burdensome emissions-reduction policies over the next decade or so, though very stringent policies may be optimal thereafter.[4] The basic argument is that to a first approximation, damages depend on long-run cumulative emissions. In the future we will know more about the consequences of our actions; we will have developed cheaper abatement methods; we will have had time to invest to prepare for their use;

we will be wealthier; and we will have higher greenhouse-gas emissions. This is of course not an argument for doing nothing today; and in particular it is not an argument against developing technologies useful in abating greenhouse-gas emissions or in adapting to climate change. But it is an argument for spending less to reduce current emissions than would be optimal if the world had to make a once-and-for-all policy choice based only on current knowledge.

Fifth, any serious program to control global emissions is almost certain to involve substantial (implicit or explicit) international transfers, the pattern of which may change over time. As the report (IPCC 1996, Section 2.4.2, 71) puts it, "International transfers, in one form or another, are likely to serve as both the building blocks of globally optimal action and the cement of global cooperation." This reflects international differences in marginal costs of abatement, with emissions reductions relative to baseline typically cheaper in poor than in rich countries (see Section 9.2.5.1, 335–43), as well as in willingness to pay for greenhouse-gas abatement. In the latter connection, it is important to recognize that the identities of rich and poor nations will likely change, along with patterns of social and political differences, in ways that are difficult to foresee. Only a few decades ago Argentina and the United Kingdom were among the very richest nations, Korea was a dreadfully poor Japanese colony, and the Soviet Union was a rapidly growing Stalinist superpower.

Finally, whatever the merits of the case for doing so, there is currently little political support for devoting substantial resources to this issue, and there is no obvious reason to expect this to change any time soon. In the United States, neither the Bush nor the Clinton administrations have gone beyond research and voluntary measures. While some other OECD nations have done more, and compulsory measures are under active diplomatic discussion as this is written, it is fair to say that no government outside Scandinavia has yet imposed burdensome restrictions on its own citizens in the name of climate change. Moreover, none have shown any serious interest in financing the massive North-South transfers that are likely to be necessary for a globally affordable transformation of the world's energy system.[5] Poor nations, of course, generally refuse to allocate any of their own resources to the climate problem, in part because they generally have trouble finding the resources to solve environmental and other problems that are literally killing their citizens every day.

Climate change is a difficult issue for the world's political system. There is no world government, so individual nations will participate in climate-related activities, including emissions control, only if they believe that the tangible and intangible benefits to them of doing so exceed the costs. Because the problem is global, unilateral national emissions reductions will generally involve costs today and at most minuscule national benefits ever. Because nations distrust each other and a round of broken emissions stabilization pledges will not help this,[6] governments may be reluctant to spend resources to honor multilateral agreements. Because uncertainty is so high, of course, there is no guarantee that even global emissions reductions will yield any benefits. Under these conditions, refusal to go beyond symbolic actions is not surprising. As Skolnikoff (1990, 78) puts it, " ...outside the security sector, policy processes confronting issues with substantial uncertainty do not normally yield policy that has high economic or social costs." Moreover, only the environmental movement is pressing for action on climate change, and many other interest groups are opposed.

All this seems to have clear implications for the near-term policy agenda. As uncertainties are resolved and new ones are discovered, the perceived threat of climate change will almost certainly change in importance over the next few decades. There is no guarantee that this change will be monotonic: we "learned" from models that the ozone depletion problem was not as serious as had been initially believed, before detection of the ozone hole refuted those models. A key task of current policy deliberations thus must be to seek inexpensive, politically salable actions that can be taken today to reduce the costs of substantial reductions in future emissions, should they become desirable.[7]

Central to this task must be establishment of effective institutions for policymaking,[8] as well as a policy architecture that permits efficient transitions between particular policies. When time is measured in centuries, the creation of durable institutions and frameworks seems both logically prior to and more important than choice of a particular policy program that will almost surely be viewed as too strong or too weak within a decade. Writing before Rio, Skolnikoff (1990, 92) captured the importance of process nicely:

> Stringent policies to cut emissions may be politically impossible or even inappropriate today; but if they prove to be justified in the future, it would be of enormous value to have a clearer idea of the issues at stake, the policy alternatives, and a process for rapid response.

In the face of great uncertainty, robustness and flexibility are key to minimizing expected regrets,[9] and their achievement requires attention to institutional design rather than to policy details. This is not a call to do nothing, just as a call to focus on near-term emissions reductions is not necessarily a call to take effective action. As Norwegian Prime Minister Gro Harlem Bruntland (1996) recently put it, "An ambitious short-term emissions reduction target without the introduction of long-term practical policies does not necessarily imply a commitment to a long-term global reduction strategy."

DEVELOPING INSTITUTIONS AND ARCHITECTURES

I do not think the report analyzes architectural and institutional issues in as serious and thorough a fashion as they deserve, and it thus pays insufficient attention to the development of "long-term practical policies." As Prime Minister Bruntland (1996) has said, speaking of the Rio negotiations, "We knew the basic principles on which we needed to build: cost-effectiveness, equity, joint implementation, and comprehensiveness. But not how to make them operational." In this section and the next I want to consider the report's treatment of two important operational issues: the importance of institutional and architectural design as against policy choice, and the implications of measurement and enforcement problems.

Much of the report, particularly Chapter 11, is written as if the world were facing a once-and-for-all policy choice. In this context, Chapter 11 considers command-and-control regulation, emission taxes, and tradable permit or quota regimes and comes down in favor of tradable quotas.[10] Presumably policymakers are to deduce the fairest international allocation of quotas from the analysis of equity in Chapter 3. This picture presents a number of problems.

To start with, as Carraro and Siniscalco (1993) and Heal (1994) have emphasized that, in the absence of a world government, substantive actions on climate change will be taken by sets of nations only if each nation believes it benefits on balance—taking into account international transfers and any intangible benefits from altruistic behavior. Chapter 2 of the report does make this point in passing. However, the report does not note that there is no reason to think that any relation exists between Chapter 3's principles of fairness and quota allocations that will induce widespread participation in abatement programs.[11] Even if widespread participation is not an objective (as it does not appear to be at present), a political process is necessary to allocate quotas in any tradable quota system. When the stakes are substantial, it is generally difficult to explain political outcomes using simple philosophical principles. In the United States, for instance, tradable permits to emit sulfur dioxide were allocated to electric utilities in 1990; the pattern of specific allocations cannot be explained by any simple principle or rule.[12]

The report does not recognize the importance of political decision-making in this context, so it does not consider how a political process might operate to allocate tradable emissions rights or what sorts of institutions might best facilitate its operation. This omission would not matter, of course, if the Framework Convention's Conference of Parties had already established an adequate institutional structure for this purpose. Since I believe it plainly has not done so, an issue that may be vital a few years hence has been ignored.

In fact, serious operational and political problems make it unlikely that the world will soon adopt anything like a serious (that is, expensive) global system of tradable emissions quotas or permits.[13] Thus the report's concentration on the desirable properties of such systems seems to leave concerned policymakers nothing constructive to do in the short run but struggle against barriers to their adoption. Similarly, Chapter 11 seems to suggest that the only interesting research topics relate to implementation problems of these sorts of regimes. In fact, both policymakers and researchers confront issues today that have implications both for menus of feasible future policies and for transitions to such policies. Moreover, the inevitability of multidimensional social, economic, and scientific change on the time scales involved here makes once-and-for-all adoption of any particular set of climate-related control policies inconceivable.

At the simplest level, it seems almost inevitable that the optimal stringency of emissions-control policies will change over time in response to changes in scientific knowledge and the development of new technologies. Thus, even though Chapter 11 appears almost exclusively concerned with once-and-for-all adoption of either a long-term emissions trajectory or of "hard" or "soft" abatement policies, such a decision would be both extremely unwise and almost certainly temporary. Any international climate regime that responds to new evidence and swings of opinion will change course over time, so an important near-term task is to establish institutions capable of doing this effectively and efficiently. If, for instance, it is decided (unwisely, I will argue) that the right policy architecture focuses on CO_2 emissions limits, it follows that one must initially confront the institutional/constitutional questions of how and how often such limits are to be revised.

Questions of this sort, which the report generally ignores, are far from trivial. While flexibility is a virtue, it is important to recognize that policy uncertainty inhibits desirable investment in new technologies and long-lived capital goods, so that stability is also a virtue. In the context of tradable permits, unanticipated changes in the numbers of permits issued will produce capital gains or losses for holders of previously issued permits. Similarly, unanticipated changes in fossil fuel prices alter the value of past investments in energy-producing and energy-using assets. Thus today's policy choice creates winners and losers from an array of possible future policy choices, and the effects of those interests will depend on the institutional/political structure within which future policies are chosen.

Joint implementation illustrates the likely complexity of future policy changes. When marginal abatement costs differ, global costs can in principle be reduced by international coordination of abatement policies. However, the disappointing U.S. experience with credit-based emissions-trading programs (in which actual emissions must be monitored *and* emissions, in the absence of the project at issue, must be agreed to) demonstrates that substantial potential cost reductions may go unrealized when transactions costs are high. The current version of joint implementation, "activities implemented jointly," involves both high transactions costs and "trading" in undefined property rights. It thus seems very unlikely to produce noticeable short-term economic gain.[14] On the other hand, attempts to reduce transactions costs and to clarify property rights may yield substantial long-term gains. And attempts to coordinate abatement policies may serve to increase developing-country participation in climate-related activities and to demonstrate the link between international cooperation and cost-effectiveness.

The shortcomings of current measurement technology raise similar issues. The report (Section 11.7, 429) notes that "technology for accurately monitoring many sources and sinks of greenhouse gases has not yet been developed." Indeed, David Victor (1991) has persuasively argued that it may be possible today to monitor only CO_2 emissions from fossil fuel combustion with the reliability necessary for a tradable permit system. As I discuss further below, it is hard to imagine any serious mitigation policy in which outcomes cannot be monitored. Therefore, it seems likely that the only policy of this sort that could be adopted in the near future would focus almost exclusively on CO_2 emissions from fossil fuels. But today's measurement problems are likely to be solved some day, and we may learn that CO_2 is less important relative to other greenhouse gases than we now understand.[15] The more comprehensive the coverage of trace gases in an abatement policy, all else being equal, the lower its global costs. Thus we need to establish policy architectures and institutions that permit changing the treatment accorded emissions of each of a long list of trace gasses.

Finally, it must also be possible to change the treatments accorded to different nations. It would be a great departure from history and from current growth projections if countries' relative incomes did not change markedly between now and, say, the middle of the next century.[16] Thus, to the extent fairness depends on relative incomes, burden-sharing arrangements that are fair today will surely not be fair in a few decades. Unfortunately, history, particularly the history of long-term economic forecasting, also teaches that it is essentially impossible to know how relative incomes will have changed over such a period. There seems to be widespread agreement now that China will continue to grow rapidly for some time, while Brazil's future seems somewhat more problematic. Not long ago, however, Brazil's prospects looked rosy, while China's seemed hopeless.

For these and other reasons, what is required is an institutional structure to generate and to reach just political bargains as circumstances—inevitably—change, as well as an architecture that permits rational adjustment of policy choices over time. There is, perhaps, much to be learned from the GATT (now the WTO) and perhaps from other institutions including the ILO and the OECD. The report's near-total silence on these long-term issues seems likely to reinforce the unfortunate tendency of the diplomatic process to focus on "an ambitious short-term emission reduction target" to the exclusion of "long-term practical policies."

MONITORING AND ENFORCEMENT

Much of the report is written as if there were a world government capable of levying taxes, enforcing emissions limits, and defending property rights.[17] Thus Chapter 1 (Section 1.3.4.3, 30) simply asserts that "In the absence of compulsory taxation, externalities can only be addressed with well-defined property rights … and a legal system that enforces compensation for externalities" without seriously addressing the consequences of the absence of all of these elements from the current scene. All supranational discussions of climate change occur within the framework of international law, within which compliance with treaty obligations is voluntary or, in some cases, enforced by limited sanctions. And there is no provision in the Framework Convention as it now stands for any use of sanctions to compel parties to meet their obligations. While it is conceivable that global institutions dealing with climate change could somehow come to exercise the sort of supranational authority that has been given to, say, the European Union, the enormous effort necessary to create the Union in a relatively small, culturally and economically homogeneous region indicates how very far away we are from anything like "compulsory taxation."

The problem of inducing compliance with emissions-mitigation policies without the ability to impose sanctions is raised toward the end of Chapter 11 (Section 11.6.5, 426), only to be immediately dismissed:

> Indeed, it is a fundamental norm of international law that treaties are to be obeyed, and as a rule countries do not negotiate, sign, and ratify agreements with the intention that they will not comply fully with all relevant provisions. Hence, compliance is not as great a problem as it is sometimes taken to be. More difficult are

the problems of negotiating an agreement that requires real sacri-
fices by the parties and of getting countries to sign the agreement
in the first place.

Two examples are discussed just below this asserted proposition, presumably in order to support it. Instead, they seem to cast serious doubt
on its validity.

It is first argued that widespread noncompliance with the reporting
requirements of the Montreal Protocol has arisen not from bad intentions,
"but rather because [countries] did not have the resources and technical
know-how needed to carry out their obligations." As even noneconomists
know, "I don't have the money," almost always means, "I have better
things to do with the money." If half the signatories to the Montreal Protocol are willing to claim in public that they couldn't afford to meet the
Protocol's reporting requirements, it does not take much imagination or
cynicism to predict near-universal noncompliance with a climate protocol
involving costs that are orders of magnitude larger. A history of partial
compliance with low-cost environmental treaty obligations argues that
compliance with any burdensome future climate-related agreements is
likely to be very spotty indeed.

Second, it is noted that noncompliance with certain oil pollution
treaties was solved when an equipment standard was adopted that made
monitoring easy, and it is asserted in passing that "monitoring of international agreements may be the more important problem." This assertion is
hard to dispute; one can only wish its implications had been explored.
Most international environmental agreements rely on self-reporting, and
almost none are well monitored.[18] And, as I noted above, it is at least
arguable that for technical reasons only CO_2 emissions from combustion
of fossil fuels can be reliably monitored today.

Of course, as long as there is little political support anywhere for
spending significant resources to control greenhouse-gas emissions, problems of monitoring and compliance are not a binding constraint on the
policy process. If there is no effective pressure to act, barriers to action
have no consequences. But if perceptions and the political climate
change, failure to have dealt with monitoring and enforcement problems
may suffice to block significant collective action to mitigate climate
change. After all, the compliance problem is not merely that agreements
will fall short of their stated goals, but that nations fearing noncompliance
by others will not sign agreements in the first place.[19]

Part of the solution to these problems clearly lies in research on methods of measuring greenhouse sources and sinks. If the world is to have
the option of adopting a significant, comprehensive program of emissions
mitigation in the future, nations must at least be able to monitor each

other's emissions. But an important part of the solution also lies in architectural and institutional design. Any serious abatement policy requires investing in the collection of credible, internationally comparable data on sources and sinks. This, in turn, requires an institution with technical expertise, financial resources, and some degree of independence.

More importantly, I believe that, even though the report's acceptance (particularly in Chapter 11) of the importance of setting limits on greenhouse-gas emissions in the short run reflects the current tenor of international negotiations, this acceptance is nonetheless unwise.[20] Countries can almost always plausibly blame failure to meet fixed emissions targets on unexpected fluctuations in domestic output or world markets—or on the previous government. And, if only because there is a stochastic element in economic activity, and governments do change, it is difficult to imagine an international regime imposing sanctions tough enough to serve as deterrents on the basis of past violations of emissions limits, particularly in the face of (nearly inevitable) promises to do better in the future.

ALTERNATIVE ARCHITECTURES

Because the report does not seriously consider alternative institutional paths to strict global emissions control regimes, it suggests that architectural issues of policy sequencing and dynamics are unimportant. This, in turn, tends to support simply doing the easiest tasks first. I believe that international negotiations are currently taking us down a path of this sort, with inadequate thought being given to where it is likely to lead.

The current focus of international negotiations is on achieving reductions in CO_2 emissions from fossil fuel use by industrialized (Annex 1) nations over the next few decades. In the summer of 1996, the United States joined other nations in calling for "legally binding" emissions limits.[21] The only way to guarantee that any nation's emissions do not exceed any particular limit is to use a system of tradable permits domestically. The logical next step, after limits have been set, would seem to be for Annex 1 nations to impose such systems on their domestic energy markets. What will happen thereafter will depend on how seriously nations choose to take the limits to which they have subscribed.

The policy architecture implicit in this approach may be characterized as "deep, then broad," since any serious program of emissions control must involve participation by developing as well as developed nations.[22] Unfortunately, it is not likely to be easy to broaden a geographically narrow tradable permit regime. In the first place, tradable permit regimes tend to resist policy changes of any sort, since changes impose capital gains and losses on those with long or short positions in permits. In addition, as many studies have shown, any geographically limited regime would induce investment in CO_2-intensive activities in nonparticipating nations, and the owners of those investments would be new opponents of their nations' participation. The need to obtain their assent would increase the international transfers required to broaden participation. Unfortunately, little if any attention is now being paid to the institutions necessary to effect such transfers or, more generally, to produce efficient international allocation of abatement effort. As noted above, achieving such an allocation would require moving well beyond the current pilot phase of joint implementation.

Schelling's reaction to commitments he anticipated, that rich nations would make to specific percentage reductions in emissions, points to an alternative architecture that I believe is superior on both environmental and economic grounds:[24]

> I cannot help believing that adoption of such a commitment is an indication of insincerity. A serious proposal would specify policies, like taxes, regulations, and subsidies and would specify programs (like research and development), accompanied by very uncertain estimates of their likely effects on emissions. In an international public forum, governments could be held somewhat accountable for the policies they had or had not put into effect, but probably not for the emission levels achieved (Schelling 1992, 13).

As I interpret Schelling's comments, they point to a "broad, then deep" architecture. This alternative would place less stress on near-term emissions reductions, which are of relatively little importance over the long haul, and would concentrate instead on developing institutions to ensure broad international participation in emissions abatement, which is essential to any serious effort. "Participation" would involve at least qualitatively similar abatement obligations, though it could involve substantial quantitative differences as well as a division between donor and recipient nations. "Deepening" would involve later tightening constraints on global emissions and, perhaps, developing the institutional and policy framework necessary to give teeth to "legally binding" emissions constraints, when and if participating nations make a collective decision to do this.

In order to enhance participation in a "broad, then deep" approach, I believe attention should not initially focus on actual emissions, which are affected by many factors beyond governments' control. Instead, I would follow Schelling and adopt a hybrid between tax and tradable permit regimes. This hybrid would involve international review of government actions, as do proposals for harmonized greenhouse-gas taxes, but, unlike those proposals, it would not prescribe the form of domestic emissions-control policies. Such a system would make participation more attractive by not forcing nations to choose between adopting tradable permit systems and risking involuntary violations of their treaty commitments. It would involve internationally negotiated (and thus differentiated) emissions targets, as do tradable quota or permit schemes. Negotiations would concentrate on maximizing participation at acceptable cost, not on implications of abstract notions of fairness, and targets would accordingly not be burdensome on average in the short run. In order to provide policy flexibility, nations would demonstrate compliance by showing ex ante that their targets would likely be met rather than by demonstrating ex post that they were actually met.

The general approach of concentrating on ex ante evaluation of policies rather than ex post assessment of outcomes is not common in environmental treaties, but it has been employed to good effect by the OECD and the IMF, among other international organizations.[24] It also bears some resemblance to the administration of clean air policy in the United States, which involves federal review of state implementation plans that link planned actions with achievement of air quality standards. In any case governments can more easily be held accountable for current policies than for past emissions, since the latter depend on policies in effect and random shocks occurring in the past.[25]

Key to this hybrid approach, of course, is the ability to relate a nation's current policies to its likely future net trace gas emissions. This requires developing both data sources and modeling capabilities, as both the OECD and the IMF have done. If international public opinion is to be the main enforcement agent for the foreseeable future, and I believe this is likely to be the case, then public opinion should be well and credibly informed, at the least, by able and objective audits of national emissions forecasts. Developing an international institution capable of predicting individual nations' greenhouse-gas emissions with accuracy comparable to, say, OECD predictions of national inflation rates is a difficult task, but, I would argue, an important one.

Both the "deep, then broad" and "broad, then deep" architectures imply feasible short-run agendas. The latter would build information, institutions, and international participation that have considerable insurance value, while the former would have difficulty moving beyond limited abatement efforts in a few nations. The hybrid approach described above would prepare the ground for more stringent policies, should they turn out to be justified. An important challenge would be to make this approach consistent with effective international cooperation involving both equitable burden-sharing and equalized marginal abatement costs. At the very least, it is hard to see the case for adopting a "deep, then broad" architecture based on tradable permits without any serious analysis of "broad, then deep" architectures or other alternatives.

SOME ADDITIONAL ISSUES

The report's discussion of several issues not mentioned above could have been more useful to policymakers. First, and in some ways most important, is the report's failure to point out the inconsistency between the cost-benefit approach and the objective specified in Article 2 of the Framework Convention:

> ...to achieve stabilization of greenhouse gas concentrations in the atmosphere at a level that would prevent dangerous anthropogenic interference with the climate system. Such a level should be achieved within a time frame sufficient to allow ecosystems to adapt naturally to climate change, to ensure that food production is not threatened and to enable economic development to proceed in a sustainable manner.

The first and more often quoted of these sentences presumes the existence of a threshold level of greenhouse-gas concentrations, above which lies danger and below which lies safety. It seems to me that the charge to explore the use of cost-benefit analysis to inform climate-change decision-making carries with it the requirement to consider the consistency of such analysis with the climate-change objective that has been adopted by the international community. Since I have seen nothing suggesting the existence of a meaningful concentration threshold, I suspect this emperor has no clothes. One could argue that the report says as much, implicitly, by discussing the assessment of costs and benefits elsewhere instead of the existence or measurement of thresholds; but straight talk would have done much more to elevate the level of debate.[26]

The second sentence quoted above seems to involve thresholds relating to rates of change of atmospheric concentrations. It appears to presume that ecosystems and food production can adapt without harm to rates of change below some level, while economic development will not be adversely affected if emissions and thus rates of change are (at least temporarily) above some level. For ecological and development goals to be consistent, as seems to be presumed, the first of these critical rates of change must exceed the second. Nothing I have seen in the report or elsewhere justifies any of these presumptions. If the Convention's objectives require policy debates to somehow be driven by unknown and probably imaginary thresholds, policymakers should surely be told this. The high cost of a serious mitigation program could be substantially increased if it must be erected on an unsound conceptual foundation. On the other hand, if the Convention's stated objectives were not intended to be taken seriously, IPCC Working Group I's much-publicized analysis of stabilization at alternative concentration levels was largely wasted effort.

Second, the report almost completely ignores the nontrivial scientific/economic problem of how to compare emissions of different greenhouse gases for the purpose of designing policy.[27] In the cost-benefit framework, comparisons of emissions of different gases at the margin must logically be based on discounted net damages. This basic principle implies that the Global Warming Potentials (GWPs) computed by Working Group I and endorsed by the IPCC have no logical foundation or value for policy-analytic purposes. If this principle is ever stated (or disputed, for that matter) in the report, I missed it.[28] Certainly, the report does not attempt to apply this principle. It thus ignores an economic question that is central to the design of the comprehensive, multigas policies mandated by both the Framework Convention and common sense.

Third, the report also pays insufficient attention to analytical issues raised by North-South resource transfers. At several points the report does note both the likely importance of such transfers in any substantial mitigation effort and the difficulty of effecting them; for example, Section 2.4.2, 71:

> Nevertheless the political and managerial difficulties surrounding such transfers need to be understood and respected by all parties if the process is not to collapse into an unproductive struggle over resource transfers.

Unfortunately, "the political and managerial difficulties" are not spelled out, nor are any related economic issues examined. Chapter 11 considers the potential role of carbon taxes or allocations of tradable quotas or permits in effecting transfers, but the discussion stops short of providing any useful guidance—or even indicating whether such guidance can currently be provided.

Finally, the report properly notes at several points that because of the long time intervals involved, the development of new technologies has the potential to reduce dramatically the ultimate costs of both mitigation and adaptation policies. And at several points the report goes on to argue for increased government spending for basic and near-basic research as insurance against the need to adopt stringent climate-related policies in the future. Most integrated assessments similarly conclude that near-term policies should accelerate the development of technologies that would be useful in connection with such policies.

But advances in basic research do not generate new commercial technologies without considerable additional investment, and it is not clear how, if at all, governments can usefully enhance this critical stage of the innovation process. As the report (Section 1.5.4, 37) correctly notes, "...there is a general consensus among economists that the patent system provides a better basis for financing applied research than do government grants, largely because of the difficulties government has in picking those innovations most likely to produce high returns." U.S. experience appears consistent with this consensus.[29] Thus we have a potentially important and difficult question that the report ignores: How can governments most efficiently encourage the development of new technologies that will reduce future abatement and mitigation costs? A related set of ignored questions pertains to efficient policies to enhance North-South technology transfer.

In a recent paper, Grubb, Chapuis, and Duong (1995) argue that a good way—perhaps the only good way—to encourage development of energy-saving technologies is simply to raise the price of energy. Thus they argue that recognizing induced innovation, for instance, tends to raise optimal global carbon taxes. An interesting question in this context is whether or not slightly higher energy prices tend to accelerate development of technologies useful at much higher energy prices. It seems at least plausible that slightly higher prices would call forth incremental improvements, while much higher prices would induce only investigations into radical departures from current technologies. If this is true, the insurance value of feasible induced innovation is limited, and the search for alternative, relatively efficient approaches to encouraging the development and deployment of efficient, greenhouse-friendly technologies becomes more important.

CONCLUDING REMARKS

On the whole, the IPCC Working Group III report is a very impressive document that clearly embodies many, many hours of hard and competent work. It presents much useful information on the economics of climate change, and its discussions of cost-benefit analysis and the principles of cost-effective environmental policy should be required reading for policymakers with a wide range of responsibilities. Its overall negative evaluation of traditional command-and-control approaches to environmental policy deserves to be read closely by policymakers—particularly those advocating adoption of common standards of various sorts.

On the other hand, the report does not cover in adequate depth some issues that are important to near-term decisions. Considerable attention is devoted to hypothetical once-and-for-all policy choices that are not on the table, while the longer term implications of current decisions are largely ignored. While the report does note that the stringency of optimal abatement policies is likely to vary over the next several decades, the implications for institutional and architectural choice are not explored. The report understates the importance of monitoring and enforcement problems. It is written as if a comprehensive global tradable permit regime were a live policy option instead of a possible but distant goal. It ignores the value of building global participation and climate-related institutions and accepts the myopic focus of current international negotiations on (relatively) short-term reductions in emissions from industrial nations. To be clear, all these gaps mirror shortcomings in the existing research literature and policy debates, but that does not make them less important.

ACKNOWLEDGMENTS

An earlier version of this paper was presented at the NBER Conference, "Economics and Policy Issues in Global Warming: An Assessment of the Intergovernmental Panel Report," July 23–24, 1996, in Snowmass, Colorado, and a very early version of portions of this essay was presented at an IPIECA workshop in Lisbon in November 1993. I am indebted for financial support and intellectual stimulation to the MIT Joint Program on the Science and Policy of Global Change and to Jae Edmonds, Henry Jacoby, Bill Nordhaus, Eugene Skolnikoff, David Victor, and participants in the Snowmass and Lisbon meetings for thoughtful and useful suggestions.

ENDNOTES

1. Schelling (1992) reaches broadly similar conclusions.

2. See, for instance, Alcamo and others (1995) and Schmalensee and others (1998).

3. See, for instance, Richels and Edmonds (1995) and Chapters 8 and 9 of the report. I am certainly not arguing that CO_2 should be the only focus of mitigation efforts, since it is not the only cause of the problem. However, CO_2 from fossil fuels is both the most important anthropogenic emission tending to increase radiative forcing and the source for which mitigation options are best understood; thus I believe that CO_2 is likely to be the main target of any serious mitigation policies adopted in the next few decades.

4. Most of the relevant "integrated assessment" studies, which include Manne and Richels (1992), Nordhaus (1994), and Kolstad (1993, 1994), are discussed in Chapter 10; see in particular Section 10.5.3, 368–88.

5. At the MIT Global Change Forum held in Oslo in May 1996, during a discussion of the ongoing negotiations on emissions limits for Annex I nations, it was suggested (not by the author) that perhaps negotiations should focus instead or in addition on the amount of money each rich nation would contribute to an international fund aimed at solving this problem, since any serious solution would clearly involve substantial North-South transfers. (This is consistent with Schelling's [1992, 14] suggestion that "While the developing countries are feeling their way into some common attack on their own carbon emissions, a tangible expression of their interest and an effective first step would be to establish a permanent means of funding technical aid and technology transfer for developing countries, as well as research, development, and demonstration in carbon-saving technologies suitable to those countries.") One negotiator rejected this approach out of hand, and no one else even bothered to address it.

6. I believe that there is a good chance that Skolnikoff's (1990, 91) assertion on the eve of Rio that "A premature commitment to action can pose dangers of error and backlash and can incur costs that would affect a wide variety of interests" will turn out to have been prophetic.

7. I do not mean to suggest that an exclusive focus on abatement is appropriate, though the international policy process seems to have adopted such a focus. Logically, it is also important to see what can be done today to reduce the costs of adapting to future climate change.

8. I hope it becomes clear in what follows that I am not simply calling for the adoption of a voting rule by the Framework Convention's Conference of Parties.

9. In a world of uncertainty, all policies will normally generate regrets with positive probability, since there will normally be one or more states of nature in which another policy would have been better. Policies that have positive net benefits, whether or not the climate problem turns out to be serious, have been widely described as "no regrets" policies. I believe this terminology is misleading for two reasons. First, "no regrets" policies may impose unacceptable costs on some groups, even though they have positive net benefits in aggregate. These groups may profoundly regret adoption of "no regrets" policies. Second, if the climate problem turns out to be serious and only "no regrets" policies have been adopted, many will regret that more was not done. The level of debate might have been elevated if the report had made these elementary points or at least avoided use of "no regrets."

10. The report usefully distinguishes between emissions quotas, allocated to governments and tradable only between governments, and emissions permits, initially allocated to governments but tradable between any parties to which governments allocate or sell them. This distinction is not important for my purposes, and I will refer to these regimes interchangeably.

11. On this point, see Edmonds, Wise, and Barns (1995).

12. See Joskow and Schmalensee (1998).

13. Chapter 11 of the report outlines some of these operational problems. For further discussion, see Chichilnisky and Heal (1995), Tietenberg and Victor (1992); see also Victor's (1991) argument that monitoring problems would currently limit any tradable permit scheme to CO_2 emissions from fossil fuel combustion.

14. On the U.S. experience, see Hahn and Hester (1989); for discussions of the potential of joint implementation and the steps necessary to realize it, see Torvanger and others (1994), Selrod and others (1995), and Richards (1996).

15. It is, of course, equally likely that we will learn that CO_2 is relatively more important than we currently believe.

16. For discussions of this point and its implications in the climate context, see Edmonds, Barns, and Ton (1993) and Edmonds, Wise, and Barns (1995).

17. Indeed, one sometimes gets the impression that some sort of global climate czar is implicitly assumed. Thus, in discussing the effects of future carbon taxes, the report notes (Section 1.5.5.5, 39), "If that were the only matter of concern, one could simply announce a commitment to impose such a tax sometime in the future ...." Who could the "one" in this sentence possibly be?

18. See, for instance, Ausubel and Victor (1992) and U.S. GAO (1992).

19. This is but one reason why, in a world of poorly monitored treaty compliance, it is a bit hard to take seriously the assertion quoted above that compliance is less of a problem than inducing nations to sign.

20. The report asserts (Section 11.7.3, 431) that "the main advantage" of a tradable quota scheme over a carbon tax approach is that under the former "the resulting global emissions will be known with certainty for a global agreement and, net of carbon leakage, for a nonglobal agreement." It is at least very optimistic to assume perfect compliance in the absence of sanctions, as this statement does, and it is absolutely unclear why, when incomes, technologies, abatement costs, climate processes, and impacts are uncertain, there would be some special value in making emissions certain even if one could do so. On the other hand, as the report also notes, it is very difficult to compare the net, effective carbon taxes implied by different national tax/subsidy systems, so that harmonizing carbon taxes internationally may well be impossible.

21. Statement of Hon. Timothy E. Wirth, U.S. Undersecretary of State for Global Affairs, to the Second Conference of the Parties to the Framework Convention on Climate Change, Geneva, July 17, 1996.

22. It will also ultimately be necessary to broaden emissions coverage beyond CO_2 emissions from fossil fuel use, but I believe the problem of broadening in this direction does not differ in severity between the alternative architectures considered here.

23. To be clear, I do not believe that all nations that have adopted percentage reduction commitments have necessarily done so insincerely. "A" can be an indication of "B" without their being perfectly correlated.

24. For an insightful discussion of this approach and of the importance of developing data and expertise (along with a critique of the current "targets and timetables" approach), see Victor and Salt (1995).

25. Largely for this reason I would favor an international carbon tax over a tradable permit or quota scheme, had I not been convinced by Chapter 11 that the carbon tax involves more serious implementation problems.

26. Similar problems have arisen in other contexts, of course. U.S. Clean Air legislation rests on the assumption that there are threshold atmospheric concentrations of "criteria pollutants" below which human health is protected "with an adequate margin of safety." Limitations of measurement typically make it possible to find concentrations below which no health effects have been detected and, generally with a straight face, to identify those as the sought-after thresholds, but there is little support for the idea that such thresholds actually exist. See, for example, Portney (1990, 31–36).

27. For treatments of this problem, see Hammitt and others (1996), Kandlikar (1995), Reilly and Richards (1993), and Schmalensee (1993). The last of these papers is cited in passing in Chapters 1 and 11.

28. The closest the report seems to come to such a statement is to note (Chapter 1, note 8, 41) that Schmalensee's (1993) criticisms of GWPs would be immaterial "if all greenhouse gases had the same rate of decay." But, of course, they don't.

29. For an overview in the context of energy supply technologies, see Schmalensee (1980). Few observers would claim that the U.S. government's large investment in research on those technologies has yielded competitive returns.

REFERENCES

Alcamo, J. and others. 1995. An Evaluation of the IPCC 1992 Emission Scenarios. In *Climate Change 1994*, edited by J. T. Houghton and others. Cambridge: Cambridge University Press.

Ausubel, J. H. and David G. Victor. 1992. Verification of International Environmental Agreements. *Annual Review of Energy and the Environment* 17:1–43.

Bruntland, Gro Harlem. 1996. *Burdensharing under the Climate Convention*. Remarks at the MIT/Cicero Global Change Forum, June 13, Oslo.

Carraro, Carlo and Domenico Siniscalco. 1993. Strategies for International Protection of the Environment. *Journal of Public Economics* 52:309–28.

Chichilnisky, Graciela and Geoffrey Heal. 1995. *Markets for Tradeable CO_2 Emissions Quotas*. Economics Department, Organization for Economic Cooperation and Development, Working Paper 153, OECD/GD(95)9.

Edmonds, Jae, David W. Barns, and My Ton. 1993. The Regional Costs and Benefits of Participation in Alternative Hypothetical Fossil Fuel Carbon Emissions Reduction Protocols. In *Costs, Impacts, and Benefits of CO_2 Mitigation*, edited by Y. Kaya and others. Laxenburg, Austria: International Institute for Applied Systems Analysis, CP-93-2, June.

Edmonds, Jae, Marshall Wise, and David W. Barns. 1995. The Costs and Effectiveness of Energy Agreements to Alter Trajectories of Atmospheric Carbon Dioxide Emissions. *Energy Policy* 23(April):309–95.

Grubb, Michael, Thierry Chapuis, and Minh Ha Duong. 1995. The Economics of Changing Course: Implications of Adaptability and Inertia for Optimal Climate Policy. *Energy Policy* 23(April):417–32.

Hahn, Robert W. and Gordon L. Hester. 1989. Where Did All the Markets Go? An Analysis of EPA's Emissions Trading Program. *Yale Journal on Regulation* 6:109–53.

Hammitt, James K., Atul K. Jain, John L. Adams, and Donald J. Wuebbles. 1996. A Welfare-Based Index for Assessing Environmental Effects of Greenhouse-Gas Emissions. *Nature* 381(23 May):301–3.

Heal, Geoffrey. 1994. Formation of International Environmental Agreements. In *Trade, Innovation, Environment,* edited by C. Carraro. Dordrecht: Kluwer Academic Publishers.

IPCC (Intergovernmental Panel on Climate Change). 1996. *Climate Change 1995. Economic and Social Dimensions of Climate Change: Contribution of Working Group III to the Second Assessment Report of the Intergovernmental Panel on Climate Change.* Edited by J. P. Bruce, H. Lee, and E. F. Haites. Cambridge: Cambridge University Press.

Joskow, Paul L. and Richard Schmalensee. 1998. The Political Economy of Market-Based Environmental Policy: The U.S. Acid Rain Program. *Journal of Law and Economics* 41(April):89–135.

Kandlikar, Milind. 1995. The Relative Role of Trace Gas Emissions in Greenhouse Abatement Policies. *Energy Policy* 23:879–83.

Kolstad, Charles D. 1993. Looking vs. Leaping: The Timing of CO_2 Control in the Face of Uncertainty and Learning. In *Costs, Impacts, and Benefits of CO_2 Mitigation,* edited by Y. Kaya and others. Laxenburg, Austria: International Institute for Applied Systems Analysis, CP-93–2, June.

———. 1994. *Learning and Irreversibilities in Environmental Regulation: The Case of Greenhouse Gas Emissions.* University of Illinois, Department of Economics. PERE Report No. 94–001, January.

Manne, Alan S. and Richard G. Richels. 1992. *Buying Greenhouse Insurance: The Economic Costs of CO_2 Emission Limits.* Cambridge, Massachusetts: MIT Press.

Nordhaus, William D. 1994. *Managing the Global Commons: The Economics of Climate Change.* Cambridge, Massachusetts: MIT Press.

Portney, Paul. 1990. Air Pollution Policy. In *Public Policies for Environmental Protection,* edited by P. Portney. Washington, D.C.: Resources for the Future.

Reilly, John M. and Kenneth R. Richards. 1993. Climate Change Damage and the Trade Gas Index Issue. *Environmental and Resource Economics* 3:41–61.

Richels, Richard and Jae Edmonds. 1995. The Economics of Stabilizing Atmospheric CO_2 Concentrations. *Energy Policy* 23(April):373–78.

Richards, Kenneth R. 1996. Joint Implementation in the Framework Convention on Climate Change: Opportunities and Pitfalls. In *An Economic Perspective on Climate Change Policies,* edited by C. E. Walker and others. Washington, D.C.: American Council for Capital Formation Center for Policy Research.

Schelling, Thomas C. 1992. Some Economics of Global Warming. *American Economic Review* 82(March):1–14.

Schmalensee, Richard. 1980. Appropriate Government Policy Toward Commercialization of New Energy Supply Technologies. *Energy Journal* 1(April):1–40.

———. 1993. Comparing Greenhouse Gases for Policy Purposes. *Energy Journal* 14:245–55.

Schmalensee, Richard, Thomas M. Stoker, and Ruth A. Judson. 1998. World Carbon Dioxide Emissions: 1950–2050. *Review of Economics and Statistics* 80(February):15–27.

Selrod, Rolf, Lasse Ringius, and Asbjrn Torvanger. 1995. *Joint Implementation—A Promising Mechanism for All Countries?* University of Oslo, Center for International Climate and Environmental Research, Oslo, Policy Note, January 1.

Skolnikoff, Eugene B. 1990. The Policy Gridlock on Global Warming. *Foreign Policy* 79(Summer):77–93.

Tietenberg, Tom and David G. Victor. 1992. Possible Administrative Structures and Procedures. In *Combating Global Warming*. New York: United Nations Conference on Trade and Development, December.

Torvanger, A. and others. 1994. *Joint Implementation Under the Climate Convention: Phases, Options, and Incentives.* University of Oslo, Center for International Climate and Environmental Research, Oslo, Report, November 6.

U.S. GAO (General Accounting Office). 1992. *International Agreements Are Not Well Monitored.* Washington, D.C.: GAO, GAO/RCED-92-43.

Victor, David G. 1991. Limits of Market-Based Strategies for Slowing Global Warming: The Case of Tradeable Permits. *Policy Sciences* 24:199–222.

Victor, David G. and Julian E. Salt. 1995. Keeping the Climate Treaty Relevant. *Nature* 373(26 January):280–82.

Comments

Greenhouse Policy Architectures and Institutions

R.K. Pachauri

Richard Schmalensee's analysis is incisive, well structured, and profound. However, there are some issues that stand out in his paper that deserve comment and contradiction. His chain of arguments are, no doubt, based on a set of perceptions about which some generalizations can be made right at the outset.

- It appears that Schmalensee has not fully accepted the realities and logic of the Framework Convention on Climate Change (FCCC) and that he has some serious reservations about this convention, even though he does not state them explicitly.
- There is a general neglect of the impacts of climate change in his analysis, both in terms of likely severity and geographical distribution.
- The author perhaps makes a bit too much of the need for political and institutional arrangements for climate-change mitigation and perhaps overemphasizes the costs that would have to be incurred in this context by different societies, particularly those of the Annex I countries. Thus, while commenting on the IPCC Working Group III report, he regards some of the material presented as "politically treacherous."

The specific comments I have on Schmalensee's chapter are given below.

R. K. Pachauri is director at the Tata Energy Research Institute, New Delhi.

Schmalensee expresses the view that the IPCC report pays insufficient attention to the long-term consequences of possible near-term choices. I agree with this completely but in a different context altogether. My contention is that the IPCC report in general provides inadequate coverage of the impacts of climate change. If we focus only on the mitigation side of policy options, we are obviously ignoring a large component of costs associated with climate change—its impact on a whole range of economic activities. I also agree that in this area the report fails to develop analytical points of which policymakers should be aware. For instance, policymakers should be made aware of the consequences of inaction in meeting the threat of climate change. Even though the science on this is still imperfect, the magnitude of costs would be so large that— even with low probabilities—the expected outcome should be clearly factored into their decision analysis. Schmalensee correctly points out that the relevant economic and physical processes do operate globally and over decades rather than years. This only means that the impacts of climate change should also be seen as spreading over decades and perhaps centuries. There are, of course, problems of measurement: the science of climate change is no doubt imperfect and is unlikely to reach a level of perfection that could say with any degree of certainty that a particular level of emissions and atmospheric concentration of greenhouse gases (GHGs) would result in a specific level and type of impact. With this in mind, I disagree with Schmalensee's statement that "few observers foresee substantial climate change for at least several decades, after emissions and atmospheric concentrations have increased substantially." He does not define what constitutes substantial climate change, and this is largely a subjective question. It could perhaps be argued that the submergence of the Maldive Islands as a consequence of a sea-level rise is not an indicator of substantial climate change because it would cost only a small amount to relocate the 200,000-odd citizens of that nation to another part of the world. But if economics is a science dealing with human welfare, then the cost of relocation to the human beings thus uprooted could be infinitely large.

Of course, implied in the views presented by Schmalensee is the assumption that the Maldives are not likely to be submerged within the next several decades. Admittedly, given the uncertainties in the science related to climate change, neither view can be adopted with any degree of confidence. However, considering the huge cost that would be associated with major movements of population in the event of submergence of low-lying islands and coastal lines around the world, I would prefer that we factor worst-case scenarios into our decisions—even though the probabilities attached to those scenarios may not appear high on the basis of compelling scientific evidence.

Several episodic occurrences that have taken place in recent years in different parts of the world indicate that the likely impacts of climate change are substantial (and therefore that climate change itself is substantial). But this, of course, is a matter for climate scientists to evaluate and comment on, as indeed the IPCC attempted to do. I would simply like to point out that we must consider very seriously the impacts of climate change in all their dimensions. Schmalensee himself mentions that the very survival of humankind depends on the earth's climate, and we are still not certain of the ability of human beings to adapt to climatic change within the time frames estimated.

The author mentions that reducing global CO_2 emissions substantially would require transformation of the energy systems of both developed and developing countries. This needs clarification and amplification. The FCCC does not require transformation of the energy systems in the developing countries, at least for the present. Nor can one hypothesize that reducing per capita emissions in the highly energy-intensive countries of the world constitutes anything akin to a catastrophe. After all, even among the Annex I countries, there are substantial differences in the production and use of energy. Most of Europe and Japan, at comparable income levels, use very different quantities of energy—not to mention the differences in sources of energy—compared with the United States. Much of the current divergence has come about during the last twenty years, following the first oil-price shock. In retrospect, this transformation has been generally smooth and has not caused any adverse economic effects or imposed any great hardship on the people involved. Schmalensee also mentions that the costs of the transformation would dwarf the costs of eliminating CFCs from the global economy. This would undoubtedly be so. The elimination of CFCs does not mark a major economic transformation of any kind as far as the world's economic history is concerned. It is, at the same time, an excellent example of international agreement and political action in an area wherein several forces initially opposed the initiatives required, particularly in the time frame originally envisaged.

In this context, the author's reference to the incurring of incremental costs being a bit mad from the viewpoint of poor nations with more immediate environmental problems is not valid. Poor nations are not required to do anything at this stage under the FCCC other than meet reporting requirements. Also, it needs to be emphasized that substantial joint benefits are to be gained from mitigating environmental problems. Hence, even if the FCCC does not require poor countries to directly address climate-change mitigation, solving local environmental problems in these countries would result in global benefits as well. These solutions are not likely to increase poverty, nor do they go against the preferences of people in these countries. In fact, the work of Willett Kempton and others (1994) clearly indicates that there is a larger and more committed constituency for environmental protection in developing countries than in developed countries. Kempton cites statistical evidence to support this conclusion. But, as clearly envisaged under the FCCC, the rich nations need to take action and not hide behind the argument that some time in the future we will have cheaper abatement methods. Such a view clearly ignores the cost of climate-change impacts, and also contradicts Schmalensee's earlier statement that economic and physical processes operate over decades rather than years. Past emissions of greenhouse gases will continue to work on the world's climate system for many decades, and the time for reductions in emissions and concentrations of these gases is already long overdue.

The author also quotes Skolnikoff's 1990 observation in *Foreign Policy* that, outside of the security sector, policy processes confronting issues with substantial uncertainties do not normally yield policy that has high economic or social costs. This commentary really applies, unfortunately, to the very freakish period of time that we are going through currently. Historically, at least since the Second World War, the benefits of foreign aid outside of security concerns have been amply demonstrated most obviously in the rebuilding of Europe, Japan, Korea, and many other countries. But the same benefits also derive in cases where the strengthening of democracy, local institutions, and all-around economic growth provide a far greater stabilizing influence than what goes under the garb of security-related expenditures. Often the neglect of economic development aid results in the breakdown of security and, therefore, consequent expenditures to meet security needs at a much higher level. Policy processes need to be influenced to take into account this set of observations, and certainly the intellectual community has a responsibility in this regard. It is of course perfectly correct that the implementation of short-term emissions-reduction targets without the introduction of long-term practical policies is meaningless, as Schmalensee points out by using Mrs. Brundtland's statement. Unless clear-cut policies supporting emissions reductions over a period of time are seen as serving the interests of the societies that are to take action, merely setting one-time emissions-reduction targets will appear meaningless.

Schmalensee seems to be unduly critical of the failure of the IPCC report to recognize the importance of political decisionmaking. The complexities of political decisionmaking being as high as they are, it is obvious that the report could not have done justice to this subject. Also, his criticism of the FCCC in this context is too severe because the convention, after all, provides only a framework and could not have put an institutional structure in place. This would have taken several years of negotiations, which is essentially what is happening with each Conference of the

Parties and the rest of the processes involved in the implementation of the FCCC.

The author also has problems with the current version of joint implementation, or activities implemented jointly, on the grounds that it involves high transaction costs, but he does not explain why he thinks these transaction costs are so high. Perhaps he bases his statement that any serious mitigation policy in which outcomes cannot be monitored should not be favored on this view of transaction costs. This sounds, however, as if he is linking mitigation policy to monitoring of climate change; obviously science has not reached a level where this could be done, nor is it likely to reach such a level in the foreseeable future. Next Schmalensee discusses the fact that the economic status of different countries could change, therefore rendering burden-sharing arrangements that are fair today unfair in a decade; but this argument is also somewhat misplaced. Surely, one cannot advocate inaction just because the rationale and responsibility for action may change in the future. All one needs to do is lay down the rules of the game whereby the burden of action will change based on the burden of responsibility. In the case of assigning responsibility for climate change, the future is certainly not as relevant as the past. Emissions of greenhouse gases and the concentrations in which this has resulted are functions of what has happened in the past. When the future becomes the past, then certainly the rules of the game will change. We need to lay down a framework for these rules of the game.

Schmalensee's discussion on reporting requirements also needs some comment. He refers to the fact that half the signatories of the Montreal Protocol are not able to meet the reporting requirements and cites this as an indicator of likely noncompliance with a climate protocol. The reality is that, in the case of the Montreal Protocol, a large number of countries have literally hundreds of thousands of units handling CFCs that are totally outside the formal system. These are small garage and pavement units, repairing refrigerators and air conditioners and filling them with CFCs. To construct an information system that is able to cover all these would not only take substantial resources but also many years. In the case of developing countries, such an outcome cannot be wished into place. Undoubtedly, there would be similar problems with the burning of biomass and emissions therefrom in many parts of the world. Thus, compliance with information and reporting requirements will take time and cannot be up to the requisite standard to start with. However, a large segment of greenhouse gases and their emission lie within the possibilities of accurate reporting. There would, of course, be significant areas where reporting would be difficult simply because one cannot measure, for instance, how much methane is emitted by each head of cattle. But, this is not adequate grounds for lack of reporting or inaction at least in those areas where there are no major hurdles or obstructions to action. Similar questions and doubts were raised when the Montreal Protocol was developed, but barring a few minor distortions and violations, by and large the implementation of the protocol has proceeded as intended.

I also disagree with Schmalensee's opinion that we may learn that CO_2 is less important, relative to other greenhouse gases, than we now assume. There is almost a century of research that clearly establishes the greenhouse effect as resulting from high concentrations of CO_2. There is no doubt a significant, even vast, area of measurements and monitoring of greenhouse gases where our knowledge is not perfect and our instruments for measurement even less so; but CO_2 emissions lends itself to vigorous, measurable, and technologically feasible actions for mitigation. We cannot fail to take these actions just because of a vague hope that other gases and emissions may be more important to the solution of the problem than CO_2. Nor can we, as mentioned earlier, ignore the impacts of climate change, to which no one, I am sure, can dispute on any scientific basis that CO_2 emissions and concentrations are a major contributor.

Schmalensee's view that any serious program of emissions control must involve participation by developing as well as developed nations is valid and appropriate. However, participation has to be along the lines of differentiated responsibility as clearly laid down in the FCCC. To the extent that actions can be taken with mutual benefits to parties in developed and developing countries, as discussed, for instance, in the provisions related to AIJ (actions implemented jointly) projects, these should be encouraged. If the author finds transaction costs of AIJ to be excessive, then we need to address how these costs can be brought down. Nothing in the FCCC defines or requires that these costs be incurred. The convention essentially leaves AIJ as nothing more than a possibility, and it is the process unfolding after Rio that has come to grips with how this possibility can be translated into action.

In conclusion I would say that the architectures and institutions, which Schmalensee has focused on so effectively, will not emerge unless the concept—the statement and vision embedded in the building he has in mind—is clearly defined and accepted. The FCCC and the IPCC are merely attempts to make that statement, and influential analysts like Professor Schmalensee must help to refine it. Only then can the related architectures and institutions follow.

REFERENCES

Brechin, Steven R., and Willett Kempton. (1994). Global Environmentalism: A Challenge to the Postmaterialism Thesis? *Social Science Quarterly* 75(June).

PART 2

Specific Climate-Change Policy Issues

6

The Costs of
Greenhouse-Gas Abatement

Tom Kram

INTRODUCTION AND BACKGROUND

What are the costs of reducing greenhouse-gas (GHG) emissions? This paper provides background information on this issue for a discussion of Chapters 8 and 9 of the Working Group III contribution to the IPCC assessment report (IPCC 1996) hereafter referred to as "the WGIII report."

The WGIII report is the result of a substantial effort, carried out by a large number of experts. The difficulties of bringing such an undertaking to a satisfactory end cannot be overestimated, in particular as the issues at stake are not only extremely complex and multifaceted, but are also the subject of fierce debate with potentially very large policy implications in sensitive areas. Any comments made in this paper should be read with these observations in mind.

The first section of this paper briefly discusses the contents of Chapter 8, Estimating the Costs of Mitigating Greenhouse Gases, and Chapter 9, A Review of Mitigation Cost Studies. The second section, What Did WGIII Achieve? highlights strengths and weaknesses of the analyses in these two IPCC chapters, focusing on relevant and important questions and the degree to which the WGIII report answers them. The final section, Some Underemphasized Features, proposes some areas for further investigation, with special emphasis on building bridges between "top-down" and "bottom-up" assessments. Some examples of insights gained from bottom-up studies are given to illustrate issues on which this type of study sheds some light and which remain largely unrecognized by top-down models. This contrast emphasizes the view that information from

TOM KRAM is project head of ETSAP at the Netherlands Energy Research Foundation ECN.

various types of modeling approaches is not only valuable in its own right but can also be exploited to contribute to a better understanding of the abatement-cost issue.

THE WGIII REPORT

The issue of mitigation costs is addressed in two chapters with fairly different focuses. In Chapter 8 of the WGIII report several conceptual and methodological aspects are introduced and discussed. WGIII's Chapter 9 presents results from recently published literature in an attempt to identify useful insights for climate policymaking.

Together the two chapters are expected to enhance the knowledge and understanding of the costs of greenhouse-gas mitigation options and policies by critically assessing the published literature on the subject. This is by no means an easy task, as a vast number of studies are carried out with many very different scopes, approaches, methodologies, and assumptions. Understandably, the results reported also vary widely, and one of the challenges is to understand the primary causes for these differences.

In simple terms, the two chapters should seek to address the following core questions:

- How costly is reduction of GHG emissions?
- How uncertain are the cost estimates?
- What differences across countries can be identified?

WGIII Chapter 8, Estimating the Costs of Mitigating Greenhouse Gases

Economic-cost estimates have been integral to the climate-change debate since the issue reached the policy agenda. Advocates of early and drastic mitigation action, primarily concerned with very serious and irreversible climate-change impacts, can point to the low cost estimates circulating to justify drastic cutbacks in emissions with limited economic implications. At the same time supporters of less drastic action—stressing the need for economically justified responses and reinforced by uncertainties surrounding the nature of, or even the mere existence of, anthropogenic climate changes—refer to higher cost estimates. In this sense the apparent lack of consensus on mitigation cost levels can be viewed as a mixed blessing, as it continues to draw the attention of stakeholders in the debate. Of course, this observation should not discourage honest attempts to close, or at least narrow, some of the existing gaps.

The WGIII report concludes that, despite the enormous upswing in technical and economic research starting in the late 1980s, two features remain from the early days: (1) a focus on CO_2 rather than on other GHGs; and (2) wide ranges of mitigation costs being reported, including negative numbers. As to the continued focus on CO_2, in addition to the explanations given by WGIII (in themselves valid and to the point), other and more objective arguments for such a focus must be mentioned. In the first place, CO_2 emissions from energy systems are relatively well known, contrary to net CO_2 emissions from other sources like terrestrial and oceanic biospheres and to emissions of other GHGs from energy systems, other human activities, and natural processes. Adding to the uncertainty about emissions, the role of other GHGs in global warming is subject to scientific uncertainties as well. All in all, the suggestion that the focus on mitigation of energy-related CO_2 is primarily coincidental, inspired by the "existence of a strong tradition in long-term energy modeling," ignores more justifiable motives. In fact, some attempts have been made to include other GHGs in an integrated way in energy modeling comparable to those made for CO_2 (see Ybema and Okken 1993). In all likelihood comprehensive coverage of all GHGs is still some time away and will need substantial additional research and data collection at the grassroots level.

In particular, following the drafting of the U.N. Framework Convention on Climate Change (FCCC) in 1992, the demand for mitigation cost studies increased as international negotiations took a direct interest in information on the economics of climate policies. The unresolved controversy on mitigation cost levels is identified as a major obstacle to economic analyses making constructive contributions, and results are often received with skepticism. The WGIII report identifies various reasons for this lack of "success": past failures of energy forecasts (for example, most of the forecasts made in the aftermath of the oil crises in the 1970s), misunderstandings about the capabilities of models in answering questions, and early attempts to use models for purposes they were never designed for.

These observations inspire insights helpful in "avoid[ing] either uncritical acceptance or total distrust of current analyses of mitigation costs" (IPCC 1996, 268). They do so by looking at possible sources of misunderstanding and misinterpretations of results.

The WGIII report explains part of the ongoing confusion by observing that "many agencies have employed models that were not initially designed to shed light on the cost of emission reductions." This is a serious shortcoming, which could justify total disqualification of some results, but the report does not pursue the issue. Likewise, the WGIII authors imply a warning against continued misuse of models when they state that "others have evolved through several versions, further complicating the picture," and this also deserves further clarification. The WGIII report does not indicate in which direction such an evolution points, nor is evidence provided in the chapters to support the claim that "part of the apparent diversity of results is simply due to the development of research itself." Without further specifics or elaboration, such statements do little to enhance trust in or understanding of analyses of mitigation costs.

Before going into any discussion of cost levels, cost determinants, approaches, and methodological issues, the WGIII report first clarifies a number of definitional questions. The authors observe that, although mitigation costs are understood to result from moving from a certain reference situation to one with lower GHG emissions, it is crucial to note that the level against which the emission reduction is measured can be either some historic benchmark or the emission level expected in some future year in the absence of explicit climate-change policies, often referred to as the "baseline."

Moreover, the WGIII report examines the different concepts of cost, ranging from the direct costs of specific measures to overall welfare costs. The report notes that the welfare cost is of primary concern when estimating mitigation cost, but it is not currently feasible to "incorporate broader conceptions of human welfare" (IPCC 1996, 270). Thus for practical purposes the widest scope analyzed relates to impacts on economic welfare, particularly GDP and its components.

Finally, the WGIII report introduces the notion of net mitigation costs: the gross costs minus any side effects that could offset some of the gross costs (IPCC 1996, 270–71). Such side effects fall into three categories: negative cost potential, economic double dividend, and environmental double dividend. The report specifically mentions the controversy about the existence and magnitude of these side effects but, remarkably, confines the discussion to possible positive side effects, automatically implying that net costs are always equal to or lower than gross costs. However, less efficient carbon tax recycling schemes could hamper economic growth and threaten employment,[1] as could drastic redistribution of R&D funds to inefficient programs (such as, perhaps, some renewable energy technologies) at the expense of other technological research programs.

Furthermore, no guarantee can be given that GHG mitigation strategies won't have negative as well as positive environmental side effects. Detailed bottom-up studies suggest that synergies between GHG abatement and other environmental problems like acidification are often, at best, very small. Table 1 illustrates the limited overlap between broad categories of emission reduction measures and various environmental problems (Kram 1995). Clearly, such trade-offs and synergies can only be observed if the underlying assessments are sufficiently detailed to distinguish the various response choices. It is questionable whether more

Table 1. Relevance of Broad Categories of Emission Reduction Measures.

Environmental issues[a]	Climate change				
	Local air quality				
	Acidification				
Emissions[a]	SO_2	NO_x	NMHC, CO	CO_2	CH_4
Fuel switching	X	–	–	X	–[b]
Energy conservation	–	–	–	X	–
Combustion technology	–	X	X	–	X
Add-on abatement	X	X	X	–[c]	–

Notes: X = Most relevant way to reduce emissions; – = Of secondary relevance (not irrelevant) for emission reduction

a. Different gaseous emissions contribute to the three major energy-related environmental issues in varying degrees. Sulphur- and nitrogen oxides (SO_2 and NO_x) lead to acid rain; the other gases do not. Together with nonmethane hydrocarbons (NMHC) and carbon monoxide (CO), SO_2 and NO_x are also responsible for local air quality problems. Finally, while carbon dioxide (CO_2) and methane (CH_4) are the main energy-related greenhouse gases, all others also contribute—indirectly—to global warming (SO_2 in the opposite direction as a cooling agent). CO_2 and CH_4, however, do not induce acidification or local air quality problems.

b. Methane emissions upstream of end-use can differ strongly between fuel types as well as between production sites and practices—venting/flaring of methane associated with oil; underground/strip mining of coal; and so forth.

c. CO_2 capture and storage can be considered for certain applications. However, in particular for storage, the prospects are still very uncertain.

aggregated approaches can capture the interactions in a satisfactory way; and we must look to disaggregated approaches to answer these questions.

The WGIII report continues by noting that GHG emission levels and the costs of mitigating them for any country relate strongly to patterns of development and technological changes, but current data and methodologies generally do not allow for explicit treatment of critical parameters. The report identifies and discusses five considerations relevant to explaining different levels of GHG emissions between countries, beyond the obvious variations in physical characteristics: (1) technological patterns in key sectors; (2) consumption patterns; (3) geographical distribution of activities; (4) structural changes, in particular the relative role of high- and low-energy-intensity industries and services; and (5) patterns of international trade.

With regard to differences among models and their results, the WGIII report presents general methodological considerations, noting that it is extremely difficult to disentangle the reasons for different results from mitigation cost studies because too many of the key ingredients (tools, assumptions, and geographical coverage) vary among them. Nonetheless, while the authors fully acknowledge the limitations of such a simple two-way split between top-down and bottom-up studies (because merging efforts makes for less clear dividing lines and because it does not fully cover all critical issues), that remains the major distinction made in the WGIII report.

As to the model structure, the WGIII report identifies four major structural characteristics of energy/economy models (see IPCC 1996, Table 8.1, 284) as well as the kinds of policy issues these characteristics address. It notes that, whereas early representatives of top-down and bottom-up models were very much antipodes on each of these four dimensions, more recently the trend is toward less contrast between the two classes. As a result, particular models can now be located on "virtually any point on the spectrum represented by each dimension," and the report deems simpler classification schemes inadequate. Despite the observed inadequacy of the distinction, the report examines the "Top-down versus Bottom-up Controversy" in considerable depth (286–89). While such a focus can be defended because of its historical importance, it is unfortunate that in the end WGIII was unable to leave the oversimplified dichotomy behind.

Summary of WGIII Chapter 8

The summary of Chapter 8 (IPCC 1996, 267) adequately recapitulates the main messages of the various sections. Overall the findings tend to identify and emphasize the difficulties associated with almost all steps in the approaches employed to estimate mitigation costs. Considering that opposing views on many of the key issues addressed can be defended on their own merits, it is not surprising that relatively little firm guidance could be provided. One valuable message is that integration of bottom-up and top-down analyses is growing and starting to generate "cross-fertilization" benefits. At the same time, it is recognized that structural differences among models remain important and that different types of models are best suited to address different kinds of questions.

WGIII Chapter 9, A Review of Mitigation Cost Studies

Chapter 9 reviews a large number of recently published studies of mitigation costs. Despite the "widely varying and often contradictory findings," an attempt is made to identify "a number of useful insights for climate policy making" (IPCC 1996, 304).

Studies are grouped by world region, including the globe as a whole; and within each region, results of top-down studies are separated from those of bottom-up studies. The concluding review is also organized with this distinction between top-down and bottom-up. The arguments of Chapter 8, then, on the decreasing value of this dichotomy, have not been heeded in Chapter 9.

The WGIII review shows that a very large number of studies has been conducted in and for the United States, either separately or as a single region in a world model. In fact, quite a few of the results listed for other regions stem from the same world (or multiregion) studies. Information from regions outside the United States is sparse.

With regard to top-down modeling, WGIII commends the comprehensive EMF-12 exercise, with its harmonized assumptions on key inputs, as giving "a wealth of useful information for policy making." The authors observe that, despite the harmonized input assumptions, CO_2 emissions rise at quite different rates because other key assumptions specific to the models give rise to differences in overall energy intensity of GDP and in the amount of CO_2 emitted per unit of energy used. Similarly, in EMF-12, both the carbon tax rates required to achieve given emission targets and the inferred GDP losses differ significantly across models: tax rates by as much as a factor 6.5, and GDP losses by as much as 3.5 (see IPCC 1996, Table 9.4, 306). There can be little doubt that the EMF study has contributed greatly to understanding crucial elements and characteristics of different approaches. However, taking note in particular of Chapter 8's conclusion that "it is the insights generated,... not the specific numerical results..., that matter," it seems odd to display averaged results. Showing the varying results as a meaningful starting point in identifying sources for the observed differences does much more justice to the study.

Generally speaking, results from bottom-up studies, which apply detailed engineering-economic analysis to assess the cost-effectiveness of alternative technological options, seem to live up to expectation. The WGIII report notes that the much finer resolution in many bottom-up models makes coupling to macroeconomic assessments difficult. The report also mentions the inherent tendency of this approach to underestimate CO_2 reduction costs.

In the review of results from U.S. and Canadian bottom-up studies, the WGIII report concludes that, despite the significant differences in approaches and assumptions, the results are "generally rather close."

The section on top-down studies for non-U.S. OECD countries contains a statement that could easily be misinterpreted: "Nations...with plentiful natural gas supplies, large existing investments in carbon-free alternatives, and/or low energy use per unit of economic activity should see less growth in carbon emissions. Hence, less carbon will have to be removed from the energy system to meet a specific target" (IPCC 1996, 313). Although logically consistent, the statement might suggest that countries meeting these qualifications can be expected to return low mitigation costs as well. However, the opposite can easily be argued: many of the opportunities still open to countries with energy-intensive industrial structures that rely heavily on fossil energy are not available to these nations, and thus (marginal) mitigation costs will be high. A comparative assessment confirms the latter tendency (see Kram 1993).

In contrast to the United States, other OECD countries often heavily tax energy purchases already, in particular transport fuels. By implication, the incremental effect of a new CO_2 or energy tax will have a smaller percentage impact on consumer prices and will therefore likely have a smaller impact on consumption. Tax levels also vary widely across other OECD countries and across fuel types and sectors (IEA 1996). For example, the total taxes reported by the International Energy Agency (IEA) on heavy fuel oil used by industries in 1995 range from 3% to 44% of the purchase price. For regular unleaded gasoline, taxes range from 55% to 80% for other OECD countries, with 28.6% reported for the United States. Thus, the impact of taxation inspired by climate policy will vary widely among all OECD countries, not just between the United States and the others.

The WGIII report confirms that studies for economies in transition (EITs) and developing countries (DCs) are few and far between. Top-down analyses typically lump them together into groups of countries (regions) in global studies. Typical features limit the value of directly comparing the mitigation cost results for these regions with those for OECD countries. As the WGIII report confirms, in EITs and DCs bottom-up studies often indicate vast potentials for zero- (or negative-) cost emission reductions. These results reflect the current economic situation in which all kinds of (government) interventions seriously distort markets and create low efficiencies compared with OECD economies. In the short term, the key questions are how and how fast the economic structure and the associated technological developments of the EITs will change. In most EITs, many of the leading sectors have largely collapsed, leading to much lower levels of GHG emissions compared to those of 1990. Newly emerging economic activities tend to center around far less energy-intensive sectors, indicating a permanent break from historical relationships between GHG emissions and GDP. Re-industrialization leading to internationally competitive product prices will require adoption of more state-of-the-art technology and management practices, eliminating much of the current differences between EITs' economies and those of the OECD.

For developing countries, the future is uncertain with regard to economic structure and the level of technological development. Unlike the EITs, however, several key DCs show ongoing high economic growth rates (for example China, India, and Korea). So even if considerable gains in efficiency from the current, often low, standards could be made, this would soon be offset by the growth of emissions induced by rapid economic growth.

Summary of WGII Chapter 9

The summary of Chapter 9 confirms the wide ranges reported for the costs of GHG emission reductions. In addition to the difficulties encountered when analyzing industrialized countries, the WGIII report notes that "many existing models are not well suited to study the economies in transition or developing country economies" (IPCC 1996, 301). Moreover, industrialized countries are studied much more than are EITs and DCs. The Chapter 9 summary does not specifically address efforts to study benefits from international cooperation involving EITs and DCs, in which there appear to be big emissions reductions available at relatively low costs. The report does mention two elements offering substantial overall mitigation cost reductions: international cooperation (exploiting differences in marginal mitigation costs) and optimized emission time paths. Both elements rely heavily on assumed differences between OECD members, EITs, and DCs. But the three groups of countries are studied to very different degrees, and the tools used often don't suit EITs and DCs. We have only a rudimentary understanding of the abatement-cost functions in EITs and DCs, so much research remains to be done before these findings can be validated.

WHAT DID WGIII ACHIEVE?

The achievements of WGIII in the area of GHG mitigation cost are enormous. In particular, its identification of different concepts, approaches, and key assumptions along with their possible influence on the results of individual studies is of very great value for further discussions on climate-change policy.

In order to serve the needs of the negotiating process, three fundamental issues need to be clarified:

1. How costly is reduction of greenhouse-gas emissions?
2. How uncertain are the cost estimates?
3. What differences across countries can be identified?

Questions (1) and (2) relate to particular target levels for emission reductions, while answers to question (3) may guide the designing of cost-effective mitigation strategies involving some form of coordinated multilateral actions.

Relatively little can be extracted from the WGIII chapters that sheds light on these three questions. The assessment would have gained in value if more attention had been given to a number of related questions.

Remaining Questions

Confronted with an overwhelming number of mitigation cost studies that do not agree on aggregate findings, analysts considering policies and strategies to mitigate GHG emissions clearly need to understand where the different results come from and what they mean. As stated earlier, the WGIII report provides an invaluable and comprehensive overview of issues relevant to understanding why models and studies are different, what the primary scope and focus of various kinds of approaches are, and how modelers and analysts have enhanced their tools in recent years. However, the report would have benefited from further reflection on four questions.

What Is the Scope and Relevance of the Mitigation Cost Estimates? Mitigation costs are distinguished at different levels, ranging from the most detailed "engineering" level dealing with direct costs of individual technical measures to the most general "welfare costs." The report gives much attention to explaining why and how costs differ at these different levels, closely related to the choice of models suitable to assess them, but hardly addresses in any systematic way the scope and relevance of the various approaches at different levels of generality. For example, a utility operator confronted with the need to acquire new generation equipment will look more at direct costs, trying to hedge against risks of new regulations or tax reforms. Despite its inherent shortcomings in accounting for economic feedbacks, the ability to identify cost-effective "benchmarks" should not be undervalued. In this light, surprisingly, one of the more obvious relevant feedback mechanisms—the rebound effect—is missing in the WGIII report. Traditional engineering models typically fail to take this effect into account, but more recent hybrid approaches look especially well suited to explore relations between the availability of technical mitigation options and demands for energy services associated with economic activity levels (Nyström 1995; Musters 1995).

From another perspective, the ongoing FCCC negotiating process is essentially entrusted to national delegations. The current agenda is only slowly moving toward addressing agreements extending beyond the years 2000 to 2010. Besides providing some feel for the broader context in the longer term, multiregional and long-term assessments (encompassing timing and regional distribution of mitigation objectives) will barely be of immediate interest here. No one challenges the relevance of looking further ahead, as attention to longer-term considerations is indispensable for shaping effective shorter-term policies. However, to understand and appreciate the WGIII chapters, more explicit distinctions among studies at various levels of scope and generality would have been valuable. As reflected by the WGIII report itself, finding adequate and practical ways to incorporate long-term aggregate findings into shorter-term decision-making processes is far from straightforward, which in turn highlights the need for more guidance on the relevance of individual results for specific analytical purposes.

What Can Be Learned from Each Type of Analysis? Initially the debate on mitigation costs centered around the marked difference in mitigation cost levels between early top-down and bottom-up assessments; indeed this difference was often labeled "technological optimism" versus "economic pessimism." In the meantime, as noted above, this simplified dichotomy has lost much of its justification as an ongoing learning process has gradually removed many of the differences between the two approaches. Still it should not be forgotten that different types of assessments really aim to answer different questions or to answer different parts of the same question.

Technology-rich energy-sector models can indicate the additional capital outlays associated with less-GHG-intensive technologies and energy carriers, and they can point to shifts in fuel quantities by type and source. The detailed results can identify promising technology and energy resource options, thereby indicating how to adjust the economy to meet emission targets or how the attractiveness of individual options changes as a result of policy interventions. Impacts on prices of energy delivered and energy services provided can be estimated by sector and end-use application. Depending upon the scope and purpose of the assessment, costs reported can reflect the perspective of energy consumers or producers, or the net costs at the national level.

On the other hand, more-aggregated models with a wider scope can estimate the overall impacts of changes induced by GHG mitigation strategies, including those involving GDP and its components.

While substantial efforts are currently being devoted to redressing shortcomings in both kinds of models—with each type adopting features from the other, thereby creating some sort of hybrid tool—the result will almost inevitably be a compromise. For the foreseeable future, a role will remain for the different models and approaches, each of which is best suited to address specific elements of greenhouse-gas-mitigation strategies and their costs.

What Minimum Requirements Should Studies Meet to Yield Credible Results? The WGIII report observes that, in the early days of mitigation cost assessments, the models employed were neither designed for nor adaptable to the task. Similarly, the results derived sometimes depended critically upon oversimplified and poorly documented assumptions. For example, some so-called demand-side assessments estimated enormous future energy-saving potentials at low or negative cost simply by extrapolating the impact of using today's best available technology to replace currently dominant technologies in current markets. One far-fetched example claims that future electricity demands for lighting can be cut by 80% by utilizing compact fluorescent light bulbs.

At the other end of the spectrum, highly aggregated top-down assessments giving insufficient consideration to the potential for technological and structural changes in energy supply, conversion, and end-use—particularly assessments without a firm empirical base—can hardly be expected to properly determine mitigation costs.

Perhaps drawing up checklists for studies that allow readers to separate the "good" from the "bad" would be going too far. Clearly a detailed screening of each and every study, and the model(s) and assumptions adopted, was not feasible within the IPCC process. Nonetheless, Chapter 9 would be much more valuable if it had laid out guidelines that helped readers to judge the quality of the inputs, methods, and outputs of the different models.

How Could Information from Analyses at Certain Levels Be Best Used as Input for Assessments at Other Levels? Despite the often repeated conclusion that no single model or assessment can provide all the answers with respect to GHG mitigation costs in a satisfactory way, relatively little progress seems to have been made in combining different approaches—for example, more detailed energy-sector models with more comprehensive economic frameworks. Incompatibility of sectoral aggregation, overlapping system boundaries, differing cost concepts, and other factors certainly complicate such undertakings. Nonetheless, the generally acknowledged strengths and weaknesses of each approach continue to attract some practitioners. The tendency, though, is more toward hybrid models, which aim to alleviate observed shortcomings by stretching the traditional boundaries. This development is still in its early days, and the WGIII report does not explicitly present any results from hybrid models. In this light, WGIII might have devoted more attention to side-by-side comparisons of studies at different levels, with the aim of identifying which kinds of assessments offer favorable perspectives for enhanced insights from hybridization. The following section presents some of these missed opportunities in more detail.

SOME UNDEREMPHASIZED FEATURES

The two chapters of the WGIII report discussed here agree that both bottom-up and top-down assessments are needed to analyze costs of abatement. However, in several instances the authors observe that this dichotomy, initially justified by the very different cost levels reported by each, is becoming less relevant. It is therefore interesting to investigate what elements from the two approaches can provide insights into key factors governing the level of mitigation cost, for example, the role of baseline efficiency.

Two observations can be made about efficiency. First, the lower the efficiency in the baseline, the higher the energy consumption and GHG emissions at a given level of economic activity (see Figure 1). Second, this lower baseline efficiency leaves "room" for relatively low-cost mitigation measures if we assume that we can remove barriers to more efficient energy technologies at modest costs to the economy. On the other hand, if we assume more efficiency in the baseline, the cost of lowering GHG emissions will tend to be higher. If the comparison is not with the lower baseline in some future year, but instead with some fixed, historic emission level, less reduction would be required to arrive at the same target.

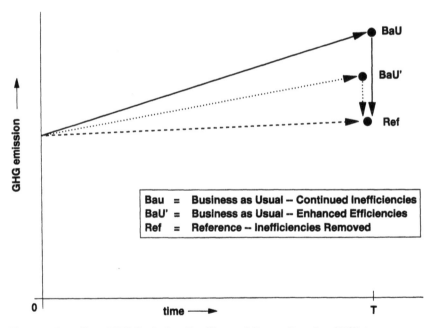

Figure 1. Baseline GHG Emission Profiles and Future Levels of Efficiency.

Note: If observed inefficiencies in the system in the initial year continue (Business as Usual, or BaU), the emission level in year T is shaped primarily by the increase in economic activity, corrected for autonomous developments and reactions to prices. In contrast, when all inefficiencies are assumed to be removed in future, the optimized system (Ref) emits far less CO_2. Various intermediate levels may be assumed where some inefficiencies are removed while others persist. This implies that the starting point for assessing reduction strategies in year T (the baseline) can vary considerably. To arrive at a target level, the highest projection (BaU) will require more emission reductions but, typically, at lower average cost and similar marginal cost than the more efficient cases (BaU' and Ref).

The unresolved issue of the existence and magnitude of zero- or neg-ative-cost emission reductions closely relates to the baseline-efficiency discussion. Figure 2 illustrates this point, building on the example pro-vided by Figure 8.2 in the WGIII report (IPCC 1996, 271). The ultimately efficient reference ("Ref") case in Figure 2 represents the borderline pro-posed in WGIII Figure 8.2; the inefficient Business-as-Usual ("BaU") case, the point O below the borderline in the same figure. For example, we can estimate the mitigation cost curve starting from the Ref point by using a least-cost optimization model, while the BaU curve might be generated by a behaviorally oriented model employing empirical price elasticities. Numerous studies have generated both types of curves. Only if we assume we can remove inefficiencies at negligible cost—that is, the mar-ket is imperfect—can a shift from the BaU curve to the Ref curve be imag-ined. Using the least-cost benchmark represented by the Ref curve might be a useful way to estimate emission reduction/mitigation cost trade-off curves instead of trying to come to grips with such poorly understood phenomena as market imperfections.

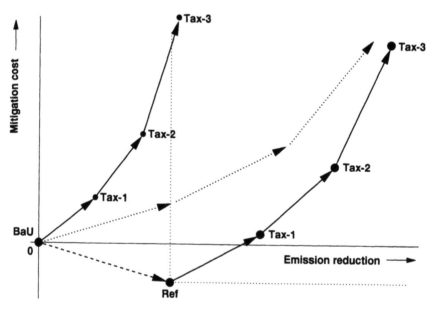

Figure 2. Cost of GHG Mitigation at Various Levels of Inefficiency.

Note: This example assumes that some negative cost potential exists, illustrated by the position of the Ref case below the horizontal axis. Starting from that point, emission-reduction cost curves can be identified by, for example, employing technological optimization models. Alternatively, such curves can be generated starting from the inefficient BaU case. If a constant level of inefficiency is assumed, this curve will be steeper than the one for the Ref case. Strategies that are more cost effective (the curve formed by the dashed arrows) can only be imagined if moving from the BaU curve to the Ref curve is thought to be feasible.

"Dynamic" Reference

One interesting question associated with the existence and magnitude of negative-cost CO_2 mitigation and the competitiveness of further energy-saving measures is whether the reference path is appropriately chosen.

As an example, look at household refrigerators, for which substantial efficiency improvements have been identified as both technically feasible and relatively inexpensive once adopted in mass production. Potential improvements include the following: Increasing the thickness of foam insulation from three to five centimeters would reduce electricity consumption by around 30% at very limited extra direct cost (not valuing the indirect cost to users resulting from loss of useful storage space inherent in maintaining fixed, standard outside dimensions). Alternatively, advanced insulation panels currently under development are expected to be more expensive but would also save more electricity (50% or more) and would not interfere with useful storage space. Finally, the cooling cycle could be improved by installing more efficient heat exchangers (estimated to yield a 15% to 20% electricity savings) and compressors (an additional 10% savings). All together, eight different refrigerator types were assessed, the most efficient one using almost 70% less electricity than the current standard (Okken and others 1994). Using the current standard (#1) as the reference, the solid triangles in Figure 3 display the additional costs measured against the savings achieved by each type. The resulting curve suggests a substantial savings potential at a cost level comparable to that of delivering baseload electricity. At least up to alternative #4, the cost of saving electricity is less than the cost of producing it. Hence, assuming a reasonably functioning market (or a successful demand-side management program), refrigerator #4 will effectively become the new standard and serve as the new reference against which to measure further savings. This situation is depicted in Figure 3 by the solid circles, showing that further electricity savings are very expensive to achieve.

Clearly, this phenomenon is not unique to residential refrigerators but holds in principle for all mutually exclusive energy devices competing for a single market. The example illustrates that interpretation of technologically based cost curves must be done with great care, as misunderstandings can easily arise. Static assumptions are unreliable guides for essentially dynamic developments, so assessment frameworks should recognize this phenomenon of dynamic references.

Best vs. Worst or Best vs. Second-Best?

Directly following from the refrigerator example is the question of why technology-rich, bottom-up models sometimes report very high marginal

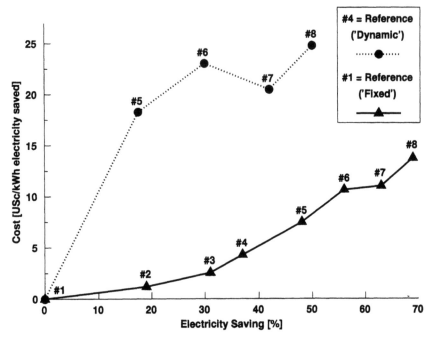

Figure 3. Electricity Saved versus Saving Cost for Residential Refrigerators.

Note: Starting from the current standard appliance (#1), seven technological improvement options for residential refrigerators are identified (#2 through #8), each characterized by electricity savings and additional nonenergy costs. Retaining the current standard (#1) as reference leads to relatively modest costs per kWh of electricity saved (lower line with triangles). Assuming, however, that with time the efficient #4-type refrigerator becomes the new standard, any further saving becomes uncompetitive, as illustrated by the upper, dotted line.

reduction costs. The answer must be that the detailed description of the energy system takes into account the existence of many alternatives for meeting a certain demand for energy or services. It then reports a marginal cost that reflects the cost associated with replacing one unit of output from the assumed second-best option with one unit from the best. For example, if supply from nonfossil electricity and fuels were assumed as the backstop, eventually the cost difference between the backstop and some very advanced, gas-fired cogeneration plant would determine the marginal reduction cost. In less detailed representations of the energy system, such as long-term top-down models sometimes use, the nonfossil backstop option typically competes with relatively expensive and carbon-intensive fossil backstops. As a result, these models obtain a bigger emission reduction at a lower incremental cost. Ultimately, all energy demands would have to be covered by either a fossil or a nonfossil backstop, and average mitigation costs would come very close to the marginal costs. In contrast, technology-rich models used to assess the medium-term (not extending beyond the first half of the next century, when concrete technological assumptions become increasingly unreliable) will return very steep marginal cost curves with comparatively lower average costs.

The Rebound Effect

Another phenomenon associated with cost-effective energy-saving measures is the so-called rebound effect. In simple terms it implies that deployment of cost-effective measures reduces the price of the energy service delivered, thereby inducing an increased demand for that service. Practical examples of demand increases following deployment of energy-saving devices are known from campaigns to promote efficient light bulbs (for illuminating formerly dark areas or leaving switches on regardless of immediate needs) and improved home insulation, both of which allowed for higher comfort levels rather than reduced fuel consumption. In addition to the direct feedbacks on the services provided by the devices themselves, net monetary saving tends to be spent again on other activities, giving rise to direct or indirect energy consumption. As a result, a larger or smaller part of the direct energy saving accomplished by the conservation measure will be offset.

The rebound effect has typically been overlooked by technology-oriented, bottom-up models, which simply assume that demands for energy services are exogenous. Although top-down approaches do rely on price-sensitive demands, suggesting that rebounds are automatically accounted for, it is questionable whether the lower level of detail in such models allows for a satisfactory representation. Exercises with more recent hybrid models better illustrate the rebound effect, indicating that an increase in demand could offset some 20% to 40% of expected technical energy savings (Nyström 1995; Musters 1995). Though the rebound effect could very well explain part of the discrepancy between strictly bottom-up and top-down approaches, the WGIII report does not explicitly mention it.

Capital Intensity of Mitigation Costs

The WGIII report rightfully notes that "a sectoral optimization of a very capital-intensive sector may not give results consistent with some form of macroeconomic optimum," recognizing a possible "shortage of capital" (IPCC 1996, 270). It is worth noting that the issue of capital intensiveness is usually raised by GHG (especially CO_2) mitigation options in the energy sector because these primarily imply a substitution of capital outlays for fuel expenditures. Closer examination of findings from engineering models can help determine the size of the problem.

Hence, in addition to looking at how total costs develop as we move to lower CO_2 emission levels, we should also monitor how the different cost components (capital, fuels, and other costs) change. Typically, capital costs show the largest increase. More energy-saving and higher-efficiency equipment enters the system, typically requiring extra capital to cut back fossil fuel use and costs. Nuclear power and renewable resources also require higher initial costs than their fossil fuel–based competitors. On the other hand, switching from coal to natural gas or to biomass fuels may increase the average purchase price of fuels. Figure 4 illustrates that, in many instances, additional capital costs dominate the additional mitigation costs reported by applications of the MARKAL technology model in a wide variety of national energy systems (Kram 1993).[2] In the majority of model runs, the increase in annualized capital costs is close to the net total cost increase (close to the solid line in figure 4) or exceeds the latter

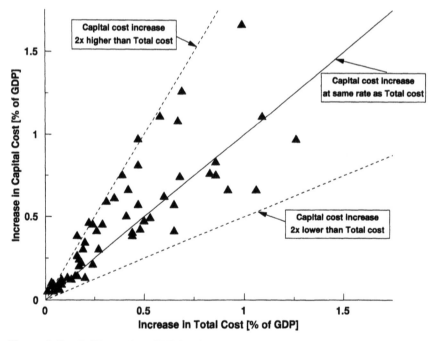

Figure 4. Capital Intensity of Mitigation Cost: Indications from MARKAL Model Runs.

Note: CO_2-mitigation measures increase the total cost of energy supply. In addition, mitigation measures like energy conservation and increased reliance on renewables and nuclear power are typically more capital intensive than the fossil fuel-based technologies they replace. In many cases the increase in annualized capital costs observed in technology models equals or even exceeds the increase in total costs (the balance of capital costs, fuel costs, and other operational costs).

by a rate of up to two (represented by the top dashed line). In all cases the extra capital cost accounts for more than 50% of the total net mitigation costs (the ratio denoted by the bottom dashed line). The issue of capital intensity and possible feedbacks from macroeconomic relationships and constraints illustrates one area of particular relevance for marrying top-down and bottom-up models.

Mitigation Cost Curves and Options Considered

"Many roads lead to Rome" when it comes to technological choices to reduce CO_2 emissions of future energy systems. In particular, when more drastic emission reductions are requested, different orientations with regard to major shifts in energy supply structures can be contemplated. Whether or not we seriously consider specific types of future options depends not only upon physical constraints (resource base, climate) and on the assumed technical and cost developments, but also on societal and political considerations. As an example, four contrasting "blueprints" for the Netherlands were drafted. Each blueprint relied strongly on one group of mitigation options, which was associated with specific characteristics making it more or less attractive from different societal and political points of view.

The "CO_2 Removal" blueprint assumed a continued reliance on fossil fuels, but the fuels were to be increasingly supplied to advanced conversion technologies with CO_2 removal. The captured CO_2 was to be stored underground in depleted gas fields and aquifers. In the "Nuclear" blueprint, the ban on the construction of new nuclear power plants, assumed in all other blueprints in light of the current low level of public acceptance, was lifted in 2005 when new reactor designs would presumably become available. "Renewables" assumed a much larger share of the Netherlands' surface area would become available for renewable energy supply (wind, solar photovoltaic, and biomass farming) than did all other blueprints. Renewables also allowed the importing of energy generated abroad from renewable resources (hydropower, photovoltaic cells in southern Europe/northern Africa, wood, and methanol). "Conservation" considered none of the three major supply options but instead adopted much more extreme assumptions on energy-saving equipment, including intermodal transport shifts (from passenger car to rail, from truck to ship).

The results in Figure 5 show that the least-cost mitigation cost curves estimated for each of the four blueprints differ sharply after an initial stage. Even without considering extensive macroeconomic feedbacks, it can be safely assumed that net economic costs of each of the four strategies differ significantly.

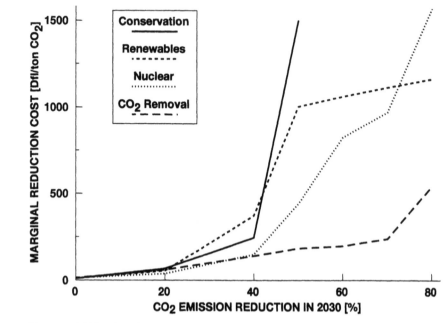

Figure 5. Mitigation Cost Curves under Four Technology Orientations for the Netherlands.

The marginal cost associated with achieving a certain level of CO_2-emission reduction depends critically upon the assumptions adopted for future technical viability and societal acceptance of key groups of options. Uncertainties in these areas add to those arising from cost and performance characteristics estimated for the options, and these additional uncertainties deserve careful consideration.

Bottom-up assessments can incorporate and document such explicit choices, thereby offering more transparency to the assessment of alternative strategies and to the sensitivities of mitigation cost estimates. The extent to which more-aggregated, top-down models currently allow us to capture choices that lie outside the traditional boundaries of economic research but have potentially serious impacts on the economics of mitigation is questionable. Harmonized assumptions are crucially important when comparing mitigation cost estimates for regions or countries and thus should extend beyond the obvious items like discount rates, prices of traded fuels, and technology characteristics. Also, to understand and appreciate differences in mitigation costs for specific countries or regions, the inclusion of assumptions on selected or excluded options is indispensable. Finally, the "value" of specific groups of mitigation options can be estimated by including and excluding them in the assessment, provided the structure of the analytical framework is sufficiently detailed.

Endogenized Learning

The rate of technological change is increasingly recognized as a decisive factor in establishing baseline scenarios. Acknowledging that demands for new technologies will not only be reinforced but also redirected in anticipation of the lowering of GHG emissions, it becomes obvious that this issue is also crucial in forecasting the economic costs of mitigation. As a rule, technological progress is assumed to result from autonomous developments, taken as exogenous inputs into scenario assessments. From that perspective, technological enhancements can be viewed as a "supply push."

Some models assume the rate of technological change to be a function of the level of R&D spending, thus making it a policy option to set aside a smaller or larger slice of the total budget (or GDP at the national level) to achieve enhanced long-run productivity. Typically, the R&D-productivity links are made at a high level of aggregation, leading to an overall, across-the-board increase in productivity from rising R&D expenditures.

Recently attention has been given to more detailed investigations of the processes driving development of specific technologies, referred to as "technology dynamics." Consideration of the conditions under which technological developments are fostered indicates that endogenizing technological change is a fruitful technique for developing new scenarios. In the case of GHG mitigation cost assessment, some form of endogenized learning may accelerate technological development of specific classes of low-GHG-emission options. In its elementary form, endogenized learning can be represented as cost and/or productivity enhancements relative to accumulated experience or levels of annual production. Hence, more-promising technologies develop faster than others, thereby capturing a larger market share, which in turn accelerates their development.

Several experiments with energy models featuring some form of endogenized learning show considerable promise (Messner 1995; Mattsson and Wene 1996). In the absence of technological learning functions, future costs and performance parameters of emerging technologies are assumed to develop along some exogenous time path regardless of the level of deployment. With technological learning, two improvements are observed: (1) new options penetrate the market more gradually; and (2) their penetration rates vary with changing market conditions such as GHG emission-reduction requirements. It must be noted that research on technology dynamics suggests that very complex processes are at work and that many factors play a role in determining which technologies emerge successfully. Ongoing research already provides insights and guidelines for the estimation of endogenous learning functions and for more appropriate concepts for endogenized learning than the relatively simple approaches used today.

SUMMARY

In summary, I would reiterate the major points of this review as follows:

First, Chapters 8 and 9 of the WGIII report contain a wealth of informative material on the many issues and factors relevant to GHG mitigation cost estimates. The illustrations of results of existing studies provided in Chapter 9 add a great deal to the more conceptual and theoretical background of Chapter 8.

Second, the observed differences in purpose and structure of the models discussed add to discrepancies that arise from differences in exogenous assumptions. The authors put considerable effort into explaining the grounds for such differences but explore less thoroughly the suitability of specific approaches for specific purposes and for the credibility of the forecasts.

Third, the strict dividing line between top-down and bottom-up results in Chapter 9 contrasts with the conceptual discussions in Chapter 8. It hampers side-by-side discussion of the meaning and validity of results from different approaches; and potentially fruitful and mutually beneficial exchanges and combined assessments remain largely untouched.

Fourth, several topics relevant to mitigation costs can only be addressed appropriately with sufficiently detailed representations of energy systems, currently offered only by sectoral models. Similarly, technology-oriented approaches often ignore potentially decisive economic feedbacks (such as the impact of greatly increased capital demands) which can only be accounted for by turning to more comprehensive economic models. These issues are not adequately addressed in the chapters.

Finally, the discussion on the existence and magnitude of "no regret" potentials remains unresolved in the WGIII report. The associated issue of the level of efficiency adopted in the baseline is in itself very relevant, but it receives too much attention considering its limited impact on the costs of meeting mitigation targets relative to historic benchmark levels.

ENDNOTES

1. Incidentally, the summary of Chapter 9 does feature a similar warning (IPCC 1996, 301).

2. The results shown here reflect the accounting rules and system boundaries of the MARKAL model, in which expenditures on durable consumer goods (for example, buying more efficient refrigerators) are also included under capital. Moreover, the model covers more than the energy sector in the traditional sense by taking into account the energy-relevant parts of other sectors (agriculture, industry, households, and commercial and noncommercial services). More detailed breakdowns of expenditures by type and sector can also be extracted from MARKAL model results.

REFERENCES

IEA. 1996. *Energy Prices and Taxes, First Quarter 1996* (quarterly publication). Paris: International Energy Agency.

IPCC (Intergovernmental Panel on Climage Change). 1996. *Climate Change 1995. Economic and Social Dimensions of Climate Change and the Contribution of Working Group III to the Second Assessment Report of the IPCC.* Edited by J. P. Bruce, H. Lee, and E. F. Haites. Cambridge: Cambridge University Press.

Kram, T. 1993. National Energy Options for Reducing CO_2 Emissions. *Volume I, The International Connection.* ECN-C-93-101. The Netherlands: Petten.

———. 1995. International Energy Technology Portfolio Assessment. Paper presented at the ETSAP Seminar on the Role of Energy Technologies towards Sustainable Development, Keihanna Plaza, Kansai Science City, Japan, 16–17 October 1995. Reprint published as ECN-Rx-95-072. June. The Netherlands: Petten.

Mattsson, N. and C.-O. Wene. 1996. Assessing New Energy Technologies Using an Energy System Model with Endogenized Experience Curves. Paper presented at the International Seminar on Energy Technology Assessment: Data, Methods and Approaches, Leuven, Belgium, 6–7 May. Gothenburg, Sweden: Chalmers University of Technology.

Messner, S. 1995. *Endogenized Technical Learning in an Energy System Model* (Draft). October. Laxenburg, Austria: IIASA.

Musters, A. P .A. 1995. *The Energy-Economy-Environment Interaction and the Rebound Effect.* ECN-I-94-053. The Netherlands: Petten, May.

Nyström, I. 1995. *Improving the Specification of the Energy-Economy Link for a Systems Engineering Model: Applications for Sweden.* December. ISRN CTH-R-95/5-SE. Gothenburg, Sweden: Chalmers University of Technology,.

Okken, P. A. and others. 1994. *Energy Systems and CO_2 Constraints.* ECN-C-93-014. March. The Netherlands: Petten.

Ybema, J. R. and P. A. Okken. 1993. *Full Fuel Chains and the Basket of Greenhouse Gases.* ECN-C-93-050. December. The Netherlands: Petten.

7

The Costs of Carbon Emissions Reductions

John P. Weyant

Significant benefits may result from efforts to limit greenhouse gas emissions. However, emissions reductions will not be imposed unless those benefits are perceived to outweigh the costs of achieving them. Thus, Working Group III (WGIII) of the Intergovernmental Panel on Climate Change (IPCC) focused a good deal of attention on reviewing the available projections of the costs of reducing greenhouse gas emissions. My review of Chapters 8 and 9 of the IPCC WGIII report (IPCC 1996)—on the costs of greenhouse gas emissions reductions—is organized in four main parts: what is in the chapters; what is included that is of potential value to the policymaking and research communities; what of potential value to the policymaking and research communities is left out or incompletely described; and what areas of future research seem most likely to improve our ability to make decisions based, at least in part, on projections of the costs of carbon emissions abatement.

WHAT IS INCLUDED IN THE CHAPTERS

Chapter 8 contains a very good and comprehensive review of a very broad set of studies and related literature on the costs of reducing carbon dioxide emissions. These studies are described and analyzed by a formidable interdisciplinary team of international experts who have reviewed and integrated a very diverse set of literatures.

JOHN P. WEYANT is Professor of Engineering-Economic Systems and Operations Research, and Director of the Energy Modeling Forum at Stanford University.

The core of the chapter is a compilation and review of a large number of top-down and bottom-up studies of the United States, Canada, other OECD countries, economies in transition, plus afforestation studies and policies designed to control other greenhouse gases. Using a set of tables and charts, the authors succinctly summarize and contrast these studies.

As part of the IPCC review, the key determinants of results from top-down studies are discussed. For example, baseline emissions depend upon GDP growth, the response to fuel price changes, technology change assumptions, and assumptions about interfuel substitution. Additionally, the cost of emissions reductions depends upon baseline emissions, the price elasticity of energy demand, capital stock dynamics, and so on. In addition, the three different models in which technology assumptions are implemented within bottom-up studies are discussed: (1) the use of end-use efficiency assumptions, (2) modeling the implementation of best available technologies within a process analysis, and (3) modeling the use of best available technologies moderated by behavioral lags and various market imperfections/hidden costs.

A number of crucial determinants of the cost estimates are also reviewed, including (1) revenue recycling, (2) trade-related carbon "leakages," (3) endogenous and exogenous technical change, (4) new technology introduction and diffusion, (5) general equilibrium considerations, (6) energy versus carbon taxes, (7) capital stock turnover, (8) disequilibrium economics, (9) growth rates and discounting in evaluations of carbon sinks (trees), and (10) discounting.

Also included are good discussions of the optimal timing of emissions reductions and the benefits of international cooperation in reducing emissions. I summarize the key points from those discussions as well as the implications of the basic cost conclusions.

THE MOST SIGNIFICANT RESULTS FROM THE CHAPTERS

A number of policy-relevant insights emerge from the review of cost estimates in Chapter 9 of the IPCC Working Group III report (IPCC 1996). First, if controls are phased in slowly and least-cost emissions-reduction options are implemented, carbon emissions could be stabilized at 1990 levels over the next fifty years for a few percent of world GDP per year. The expected cost would represent about a 0.01% reduction in world economic growth over the next few decades, or about one trillion dollars per year worldwide on average.

If all world regions limit emissions to 1990 levels, Chapter 9 reports long-run global cost projections ranging from 3.5% to 5% of world economic output per year. Since countries/regions are projected to experience different population and economic growth rates, they will experience different growth rates in carbon emissions absent additional climate policy initiatives. Thus, the cost of holding emissions to 1990 levels will vary among nations/regions. As reported in the chapter, the long-run projections of the cost of stabilizing carbon emissions at 1990 levels in the United States range from 0.8% to 2.0% per year, from 1.0% to 2.5% in the other OECD countries, and from 5.0% to 13.0% in China. International emissions trading can considerably reduce the total cost of achieving the target global emissions as reported in summary point seven below.

Second, so-called "no regrets" options (that is, carbon emissions reduction measures that could be undertaken at zero or negative present value net costs) might lead to a 10% to 30% reduction in carbon emissions levels relative to projections that do not include this potential. Two important observations can be made regarding this assessment: if implemented, this level of "no regrets" options would be significant but not enough to offset the increase in emissions projected by most analysts over the next few decades; and many "top-down" models already include nonprice-induced energy intensity reductions at this level or greater in their baseline projections.

Third, the cost of reducing emissions in any one year is nonlinear with respect to the degree of control. That is, the second 10% reduction in emissions below baseline is more expensive than the first and so on. This diminishing return to additional mitigation effort is likely to be particularly significant in the first couple of decades of the emissions control program because most energy-producing and -using capital stock lasts for more than ten years. Short-run shocks to the economy can cost several times the long-run equilibrium costs because of the unemployment and inflation that can result.

Fourth, the cost of mitigation could rise substantially if policies designed to achieve them are inefficient or implemented inefficiently. Rough estimates suggest a two- to fourfold increase in the cost of an emissions-reduction program if it is implemented too rapidly, is not implemented globally, and leads to higher cost emissions-reduction measures being implemented before lower cost options. Higher cost emissions reductions can easily be implemented before lower cost ones if a system of mandatory technology choices (for example, efficiency standards or a specific control technology) is implemented. Such "command and control" regimes have frequently been implemented in response to other environmental concerns.

Fifth, the cost of mitigation will likely vary considerably across regions because of differences in economic growth rates, energy resource endowments, economic structures, infrastructure levels, levels of technological sophistication, and systems of governance. And if stabilization of emissions at some common fraction of emissions levels in some base year is mandated, differential growth rates in the expected trajectory of emissions absent policy intervention will lead to differences in abatement costs across countries/regions. These interregional differences provide the potential for international carbon emissions trading, which is discussed further below.

Sixth, climate change and the resulting damages are related to the concentration or stock of CO_2 in the atmosphere rather than the flow of emissions into the atmosphere year-by-year. This suggests the possibility of reducing emissions less now and more later as a desired CO_2 concentration target is approached. Such a strategy would lead to higher emissions and slightly higher CO_2 concentrations over the next few decades during the approach to the concentration target. It might be worth subjecting society to this additional exposure to achieve the very substantial cost savings that would result from allowing time to replace the existing energy-using and energy-consuming capital stock, allowing adequate lead time to develop new carbon-free energy supply technologies, accounting for the opportunity cost of capital invested in carbon abatement versus other societal investments, and accounting for the process of slowly removing new additions to the atmospheric stock of carbon via the carbon cycle. Most analyses of the timing issue have been done with full-employment models, which Jaccard and Montgomery (1996) report may significantly understate the short-run costs of abruptly imposing carbon emissions limits. According to results from macroeconometric simulations reported in their paper, the cost of immediately stabilizing emissions might be several times higher when the unemployment created by subjecting the macroeconomic system to sudden shocks is taken into account (see also Rose and Liu 1995). Moreover, the political system may be more sensitive to additional unemployment per se than to the induced reductions in economic output.

Seventh, the potential for international cooperation in climate-change mitigation policies is great. Regardless of who is responsible for paying for the emissions reductions deemed to be appropriate, it makes sense to make the reductions wherever it is least expensive to do so. Heterogeneity among countries in their energy resource bases, economic development profiles, tax systems, available infrastructure, and so on leads to widely varying incremental costs of emissions reductions. Implementation will surely be difficult—since each country has its own objectives based on its own assessment of the costs and benefits of climate-change policies—but the benefits are large enough that even partial coordination (say among the dozen or so major emitters) could have substantial benefits. Indeed, this is one area where negotiations are proceeding, albeit slowly, as the phrase "measures implemented jointly" is now part of the official parlance of the climate treaty implementation negotiation process. The time required to analyze and implement cost-effective cooperative measures is yet another justification for slow but steady, rather than abrupt, implementation of climate-change policies.

These key insights on the timing of emissions reductions and the value of international cooperation in reducing emissions should directly interest decisionmakers, as should the entire spectrum of issues and projected cost ranges described in the WGIII Chapters 8 and 9.

ERRORS OF OMISSION AND COMMISSION

Some gaps in the coverage of these chapters can also be identified. These may have resulted from the authors' explicit or implicit choices regarding what to emphasize, the timing of the report, or group composition and dynamics. Virtually everything mentioned in this section is addressed again in the following section on fruitful areas for further research. This overlap reflects the fact that the omissions identified here may point to gaps in the current state of the art (because they are emerging research areas) as much as gaps in the WGIII chapters.

The chapters do not delve too deeply into the reasons for the differences in the emissions projections and emissions-reduction cost projections. The authors do state that the differences in projections result more from differences in input assumptions than from differences in the methodologies per se. This is an astute observation and one that I wholeheartedly agree with. But they make scant reference to the slowly emerging literature on why the differences exist and attempts to attribute observed differences to underlying assumptions. Three fundamental approaches to bottom-up modeling are described, as well as the general approach pursued in doing top-down modeling. These descriptions can explain a great deal of what has become known as the energy-efficiency gap—the propensity for bottom-up studies to project significantly larger improvements in energy efficiency than the top-down studies.

On the other hand, the gap that remains can be substantial and can be ascribed to (1) market failures (for example, information limitations, first-cost biases, and so on), (2) hidden costs (for example, costs left out of the engineering estimates that pertain to the complexity and time required to make a particular choice; high individual discount rates resulting from the riskiness of energy-conservation investments; or different secondary features of two options, like the different styling of two automobiles), or (3) some combination of the two. The explanation for the gap probably varies from sector to sector and from end use to end use. Knowing the source of the gap is extremely policy relevant. If market failures explain the gap and can be corrected at lower cost than the benefits that would result, there is a strong incentive to adopt policies that directly address the market imperfections. If hidden costs explain the gap, then there is no incentive for any policy intervention whatsoever, and we would be better off relying on the top-down projections in planning future policy responses.

At the most fundamental level, large differences in projections of the potential for energy-efficiency improvements can sometimes be explained simply by the differences in assumptions about discount rates used by energy consumers and differences in assumptions about the rate at which the energy-consuming capital stock can be turned over. These assumptions are generally inputs to both approaches, and the resulting differences in projected energy use have nothing to do with which methodology is employed, being determined instead by what is assumed in these two areas.

Several researchers have proposed frameworks that can be used to test any particular explanation for the energy-efficiency gap (Huntington and others 1994) and have designed and implemented stylized or fully empirical tests for them (Jaffee and Stavins 1994). These include studies on the impact of uncertainty on energy-investment decisionmaking (Metcalf 1994) and empirical studies on market and nonmarket factors influencing energy decisionmaking (Jaffee and Stavins 1994). Fuel and factor feedbacks are included in top-down studies, but not bottom-up studies; and technology specifications are included in bottom-up studies, but not top-down studies. Hidden costs may be of a general-equilibrium or economywide nature.

One problem with factoring in the results of these studies is that they have generally focused only on relatively small areas of total energy use and have not always been conclusive. Nonetheless, they provide some preliminary guidance on the choice between the top-down and bottom-up styles of analysis.

At a conceptual level a number of additional areas can be identified. Do fuel and factor feedbacks (that is, pecuniary externalities) included in top-down studies but not bottom-up studies significantly affect the results obtained? How do the demand-side technology specifications included in bottom-up studies but not top-down studies affect the results obtained? And are hidden costs of a general-equilibrium or economywide nature (that is, through the costs of induced substitution among products or induced reductions in productivity growth) rather than a partial-equilibrium level? An example of the latter would be tests or sensitivity analyses on capital availability and use for emissions-reduction programs in a developing country, which the study cited in Chapter 9—by Mongia and others (1991)—suggests may be important.

The WGIII report establishes that the input assumptions are critical to the cost projections produced by both styles of analysis but offers no critique of the assumptions that have been used in specific studies or of the process that has been used to generate these assumptions. Key inputs to the top-down analyses would be (1) population growth, (2) economic growth, (3) technological change, (4) the fossil fuel resource base, (5) capital stock turnover dynamics, and (6) the discount rate. Key inputs to the bottom-up approaches include (1) end-use energy demand projections, (2) efficiency and cost projections for new and existing technologies, (3) capital stock turnover characterizations, (4) fuel price projections, (5) projections of the costs and availability of other input factors, and (6) the discount rate. As discussed more fully below, such a critique and the development of suggestions for improving the quality and level of uncertainty about key inputs to either approach appears to be a high priority area for future research.

Another area that seems overlooked in the review and the literature is some discussion of policy alternatives to a carbon tax. These will, in general, be less efficient (that is, cost more per ton of carbon emissions reductions than a pure tax on carbon emissions) but are probably more politically feasible in most countries. Thus, it would be useful to know how much more expensive they would be. In some ways this subject would be more appropriate for the report's Chapter 11 on implementation issues, but some discussion (as with the discussion of the European carbon/Btu tax that is included) would have been useful.

The chapters do a good job of summarizing the ambiguous contemporaneous state of the art in the analysis of revenue recycling and endogenous technical change. However, rapidly emerging research in those areas suggests some revision is warranted in the emphasis placed on the various factors important in studying these issues, if not in the conclusions drawn at the time of the review. At this point, it is acknowledged that the early studies on revenue recycling and endogenous technological change were too optimistic, often implying that policymakers could offset completely the cost of significant reductions in carbon emissions through fiscal and technology policies. That these policies could be so effective is now viewed as highly unlikely. These subjects are either fair game for a critique of the review or directions for future research. They fit better with the future research directions (as may many of the other "critiques" included in this section), so they are discussed below.

HIGH-PRIORITY RESEARCH

Refinements in projections of the potential costs of carbon emissions reductions will depend upon progress in research in a number of areas. Some of these areas (like technological change) involve broadening the scope of the analysis from a partial equilibrium to a general equilibrium and economy-wide—perhaps even a multisector trade—framework. Others, like the appropriate discount rate, will likely involve the development of new conceptual frameworks and/or the collection of new empirical evidence.

Some of the most important areas for additional research are (1) further consolidation of top-down and bottom-up approaches, (2) revenue recycling, (3) technological change, (4) capital availability and flows, (5) macroeconomic impacts of emissions-mitigation programs, (6) economic-growth projections, (7) fossil fuel resource-base estimates, (8) population-growth projections, and (9) discount rates. These areas of potential future research are now discussed in more detail.

Further Consolidation of Top-Down and Bottom-Up Approaches

Analysts have, in fact, used three different approaches to interpret the past evolution of energy and industrial systems and to project the future trends and responses of those trends to changes in external conditions and policy interventions. (The discussion here follows Weyant and Yanigisawa 1996 and parallels some of the exposition in IPCC 1996, Chapter 8.) For simplicity, I will refer to these three approaches as (1) the economic approach, (2) the engineering approach, and (3) the social-psychological approach. Although most current models (both top-down and bottom-up) employ elements of each of the three approaches, differences in formulation and interpretation of results stemming from the different disciplines of the researchers involved are substantial.

The economic approach treats energy like a conventional economic good. Thus, the demand for energy fuels and energy-intensive goods and services (the "dependent variables") generally relate to a set of "independent variables" like energy prices, the prices of other (nonenergy) goods, incomes, a set of technologies, and time (time trends generally serve as proxies for changes in technologies, tastes, and other factors not explicitly included). More sophisticated analyses may include the existence of energy-conservation programs and variables representing differences in regions or countries (for example, weather, transportation infrastructure, cultural norms, and so on). These analyses are fundamentally behavioral, in that they explicitly consider human responses to various stimuli. However, the behavioral specification included in the typical economic analyses is generally not nearly as rich as that included in the social-psychological approach. In addition, these economic studies sometimes include technological detail on energy-supply technologies but less frequently on energy-using technologies.

The engineering approach focuses on the technology of energy production and consumption. The key inputs to such analyses are typically demand projections for end-use energy services like residential or commercial-space heat, industrial-process heat, and personal transportation. Engineering representations (generally composed of an energy efficiency measure, an operating cost, and a capital cost for each technology) of energy-using technologies (for example, space heaters, industrial-process heating systems, and transportation systems) are included in this type of analysis. This approach generally projects the "least-cost" end-use technologies that energy consumers being considered will select for the end use in question. An advantage of this approach is that new technologies that do not show up in historical data can be introduced.

Another advantage of the engineering approach is that saturation effects in technology adoption can be explicitly considered. For example, the growth rate in air conditioning is more rapid when a particular energy economy is just starting to implement that technology than after almost everyone already owns an air conditioner. However, engineering models tend to overpredict the adoption of new technologies because they leave out various behavioral factors associated with the new-technology adoption process, as well as existing market barriers to the diffusion of these technologies. Another difficult challenge in using this type of model is in projecting the demand for energy services. Demand projections are usually tied to projections of economic activity, which are made independently of energy prices, equipment costs, or anything else. Interestingly, social-psychological researchers explicitly address both the rate of new-technology introduction and the end-use energy-projection challenges faced by those pursuing the engineering approach. One problem that remains, though, is how to predict the introduction of fundamentally new technologies.

In many ways the social-psychological approach complements the engineering approach, and its application has typically resulted in a broader framing but a narrower scope (in terms of number of energy users and types of consumers) than energy/industrial analyses completed with the economic approach. The starting point for this type of analysis is "human needs and wants." Thus, it focuses on explaining past trends in energy-service demands and projecting future trends within a framework that includes the full range of human activities, including individual as well as social influences.

Although this approach resembles the economic approach in many ways (discussed below), there are two major differences in practice. First, the number of noneconomic determinants of energy consumption (for example, attitudes towards energy use and "energy literacy") in the social-psychological approach far exceeds that in the economic approach. Stern and Oskamp (1987) propose a multidimensional view of resource psychology—one that is more interested in contexts of actions and sees attitude-behavior processes as embedded in larger systems of beliefs, events, institutions, and influence "background factors" (for example, income, education, family size, and temperature). For example, in their review and assessment of financial incentives for energy conservation investments, Stern and others (1986) identify a number of nonfinancial goals that conflict with participants' "investor role"—for example, when concerned about "comfort and/or appearance," when acting as "members of groups," and when "avoiding hassles" (that is, taking the minimum action necessary to solve immediate problems.

Second, the social-psychological approach recognizes the social character of energy consumption and, thus, the desirability of studying energy decisionmaking at the family, community, or societal level. Most economic models either adopt an individualistic decision-making framework or aggregate over large numbers of consumers in a way that does not allow for alternative specifications of the nature of the interactions among consumers.

Although most social-psychological analyses of energy use have been used to try to understand past trends in energy use and predict trends over a one- to two-decade period (say in analyzing the efficacy of energy conservation programs of various sorts), the biggest payoff from this approach may well be over the fifty- to one-hundred-year projection period of interest in the climate-change debate. Over this longer time period, attitudes, preferences, and social organizations may change (or may be caused to change) in ways that can have very significant effects on energy use and greenhouse gas emissions. This is especially true because most of the growth in emissions is projected to occur in developing countries where attitudes and social organizations are likely to change the most. Since the early 1970s, the term "life-style" has been used by energy researchers to refer to patterned differences in behavior and resulting consumption among subgroups in the society. The most cogent definition of life-style (Lutzenhiser 1992) is probably one that equates the term with the anthropological concept of "culture"—that is, the totality of practices, meanings, beliefs, and artifacts of a social group. Both marketing researchers and social scientists have contributed to life-style research (Lutzenhiser 1993). The two disciplines seem to agree that consumer subgroups ("market segments" to one, "classes" to the other) can be identified, but they take opposite analytic approaches to the problem.

Social-psychological researchers stress that consumption differences among demographic categories (age, income, ethnicity) are likely the product of underlying class and subcultural differences and constraints. Marketing researchers, on the other hand, recognize the importance of consumption style in predicting purchases and have developed an array of analytic tools to identify clusters of consumer characteristics and attitudes associated with differences in purchasing behavior. A number of residential energy market segmentation studies have been performed for utility firms, and several classification schemes of residential and commercial customer types have been proposed for targeting demand-side management (DSM) programs. For example, a study by the Electric Power Research Institute (EPRI 1990) proposed six consumer life-style groups: pleasure seekers, appearance conscious, life-style simplifiers, conservers, hassle avoiders, and emulators. Although these techniques have proven extremely useful in short-run marketing studies, over the long run they require projecting how culture and consumer classes will evolve—which is an extremely difficult task in itself. Since it may be difficult to project these kinds of trends, we may be better off using more aggregate, traditional economic analyses. However, there appear to be some fruitful opportunities for bringing elements of both the engineering and social-psychological approaches into traditional economic analyses.

Each of the three approaches to energy analysis emphasizes different dimensions of the same system of human activities. Therefore, not surprisingly, they substitute for and complement each other in a number of areas. Some of the earliest social-psychological studies of energy use followed the Arab oil embargo of 1973–74 and were designed to help explain why energy-conservation programs (for example, subsidies and tax breaks for various energy-efficiency-improving investments by homeowners) generally failed to yield as much energy conservation as projected by engineering studies (Lutzenhiser 1993). In their review of programs that provide financial incentives for investments in energy conservation, Stern and others (1986) identify a number of reasons why persons who are targeted with such incentives may fail to act in ways that engineering analyses suggest are in their self-interest. These include lack of accurate information, restricted choice, too much time and/or too much effort required (that is, high information and decisionmaking costs), lack of trust in information sources, lack of cash, and the relative invisibility of conservation impacts and preferences that go beyond the simple cost-minimizing of energy use.

In a series of studies Kempton and associates (Kempton and Montgomery 1982; Kempton 1993; Kempton and Lane 1994), using extensive interviews, further identify the nonfinancial determinants of conservation decisionmaking and test alternative hypotheses about how households actually make such decisions. Technology options available to consumers are relatively homogeneous for household appliances, suggesting that market failures explain any difference between the projections of the engineering models and the economic models. By contrast, technology options are not nearly so homogeneous for personal transport (for example, automobiles and mass transit options with many different characteristics are available), greatly complicating the assessment process. For example, Kempton and Montgomery (1982) argue that people generally understand the energy-accounting problem in terms of average rather than marginal costs, and that they estimate the energy consumption of equipment using measures such as running time or amount of human labor replaced. This necessarily limited knowledge of technology contributes to suboptimal energy choices—for example, incorrectly believing that lights (a "visible" end use) are a large contributor to total consumption. "Folk calculations" (Kempton and Montgomery 1982) also produce estimates of energy-efficiency costs that can differ significantly from expert calculations. When an expert payback estimate is short, folk methods provide similar results. But for longer term investments, folk methods overestimate the payback period by failing to take into account the possibility of changing energy prices. The social-psychological approach can provide the perspective needed to resolve the reason for the "energy-efficiency gap" because it explicitly focuses on information availability and the level of consumer sophistication. For a much more detailed discussion of these issues see Robinson and Timmerman (1991) and Robinson (1992).

Another synergy mentioned above is the possibility of using the social-psychological approach to project the end-use energy requirements needed as inputs to the engineering approach and then using the social-psychological approach to adjust the technology approach's market-penetration projections. In fact, the social-psychological approach has been used to demonstrate that short-run responses to energy-conservation programs result from behavior modification as well as improved technologies (see U.S. DOE 1992).

Another hybrid approach that has been attempted embeds technological information in an economic framework. If a stable relationship can be determined between the cost of a particular type of energy-using equipment and its user cost, this can be used in an equipment-choice model whose inputs are that relationship, fuel prices, and incomes (McFadden 1990).

As suggested by a number of observers, the future of energy use, especially analysis designed to cover the fifty- to one-hundred-year time frame required to address climate-change issues, lies in pursuing the economic, engineering, and social-psychological approaches to energy/

industrial analysis in combination and in parallel (for example, Robinson 1992). Each perspective yields important insights into what the future might or could bring. From the perspective of policy development, as many options as possible need to be identified. Stern (1986) suggests a better dialogue between economics and the other social sciences to fill in "blind spots" in economically oriented policy analysis. One barrier to that dialogue is the fact that social-psychological research on energy and industrial systems is generally done on a much more micro level than economic or engineering research. This micro-level approach prevents the development of the kind of bottom-line results that have often been the product of applied engineering and economic research. However, for precisely this reason, insights developed from social-psychological research may be more valuable to those responsible for policy implementation and to those trying to make the economic and engineering approaches more useful. For example, Jaffee and Stavins (1994) attempt to reflect the social-psychological perspective more directly in their recent economic assessment of energy-conservation programs. Also, several researchers have used Rogers' (1963) diffusion-of-innovation model (Farhar-Pilgrim and Unseld 1982; Leonard-Barton 1981; and Leonard-Barton and Rogers 1981) to show how solar technology is diffused through social networks.

In addition to their complementarity, there are interactions between the elements considered in each domain. For example, the citizens of a country or region might be willing to use a mass transit system or alternative-fuel automobile, even if it was only available at a price and convenience premium relative to conventional gasoline-powered automobiles. This option will only be available, though, if the requisite investments in R&D and infrastructure development are incurred. Kempton (1992) explicitly notes the opportunity for "behavioral science research conducted in parallel with new technology research." Another example concerns the acceptability of various policy responses—which are much more likely to be accepted if they are perceived as being fair to potential participants, which in turn probably requires good technical and economic assessments at a minimum.

Revenue Recycling

The Costs of Revenue-Neutral Carbon Taxes. An important issue of debate in climate-policy discussions has been the extent to which the costs of carbon taxes can be reduced by judicious recycling of the revenues from such taxes. Our understanding of this issue has advanced considerably in recent years. Theoretical work indicates that the gross (or nonenvironment-related) costs of carbon taxes can be significantly reduced by using the revenues to finance cuts in the marginal rates of existing income taxes, compared with returning the revenues to the economy in a lump sum. Numerical studies consistently confirm this result (see Shackleton 1996 and others; Goulder 1995; Jorgenson and Wilcoxen 1995).

A more controversial issue has been whether revenue recycling can make the gross costs of revenue-neutral carbon tax policies vanish or become negative. If it could, the revenue-neutral environmental tax would generate a "double dividend" by both (1) improving the environment and (2) reducing the costs of the tax system. Recent theoretical work on this issue tends to cast doubt on the likelihood of a double dividend (Bovenberg and de Mooij 1994; Parry 1995; Bovenberg and Goulder 1996). This work indicates that carbon taxes tend to be less efficient sources of revenue than the factor (income) taxes they would replace. The key to this result is recognizing that carbon taxes (like other taxes on intermediate inputs or consumer goods) are implicit factor taxes: their imposition raises output prices and thereby lowers the real returns to factors like labor and capital. As a result, carbon taxes cause distortions in factor markets quite similar to those posed by explicit factor taxes. However, carbon taxes also distort (ignoring the beneficial environmental implications) the choice among productive inputs and among consumer goods. This makes the carbon tax more costly than factor taxes. Hence the revenue-neutral swap of carbon taxes for factor (income) taxes involves a positive overall cost: the double dividend does not arise.

Still, the theory leaves room for the double dividend under some special circumstances. In particular, the double dividend can arise if (1) the original tax system (prior to introducing the carbon tax) is seriously inefficient along some nonenvironmental dimension (for example, if capital is highly overtaxed relative to labor) and (2) the combination of carbon tax and revenue recycling significantly reduces this inefficiency. Whether or not the double dividend will arise thus turns on some empirical issues.

Applications with numerical general-equilibrium models tend to cast doubt on the prospects for the double dividend in the United States (Bovenberg and Goulder 1997). However, the prospects could be better in other countries, especially in economies with prior subsidies to energy. Removing such subsidies is likely to lower the costs of the tax system while promoting an improvement in the environment. Energy is subsidized in a number of developing countries, and thus the costs of reducing carbon emissions may be minimal or negative in these countries. Further empirical investigations of this issue could be most valuable to policymakers.

A practical difficulty in studying revenue recycling is what assumption to make about the baseline path of the tax system over the long time frame over which projections of the costs of emission reductions are made. Since tax systems are continuously debated and revised, it is not clear whether the most inefficient taxes will be eliminated at some point independent of whether carbon-tax revenues are available to offset them as a source of revenues.

Using revenue recycling as part of a carbon emissions reduction strategy also raises a political-economic challenge. A carbon tax would provide a new source of government revenues, creating the opportunity to redo the complicated political-economic negotiation underlying the existing tax system. Consequently, rather than using the new tax revenues to reduce other more distortionary taxes, they could easily be used to support additional government spending on programs that have lower productivity than if the private sector spent the same amount of money (see Nordhaus 1993). In this case, the overall costs of the carbon tax would be considerably higher than if the revenues were used to finance cuts in distortionary taxes. Thus, to fully analyze the impact of revenue-recycling alternatives on the overall costs of carbon taxation, we need not only analyze the impact of a carbon tax but also ascertain how the government would employ the revenues from the tax.

Grandfathered Rather Than Tradeable Emissions Permits. The issue of revenue recycling is also relevant to the choice between auctioned and grandfathered tradeable-permit systems. Auctioning the permits raises revenues that can be applied to finance cuts in existing distortionary taxes. Recent work by Parry (1995) and Goulder, Parry, and Burtraw (1997) shows that this potential for revenue recycling gives auctioned permits a significant efficiency advantage relative to a system of grandfathered permits. Indeed, this recent work indicates that revenue recycling may be necessary to allow tradeable permits to generate efficiency gains. In particular, Parry, Williams, and Goulder (1996) show that a system of grandfathered tradeable permits for reducing carbon emissions cannot generate efficiency gains unless the marginal benefits from carbon abatement exceed $25 per ton.

These studies show that, like carbon taxes, a system of tradeable permits tends to raise output prices, lower real factor returns, and thereby distort factor markets. If permits are grandfathered rather than auctioned, the negative efficiency impact associated with the reduction in real factor returns can be stronger than the beneficial impact associated with a cleaner environment. In contrast, if permits are auctioned, this negative efficiency impact is offset by both the beneficial efficiency impact of a cleaner environment *and* the positive impact of using the revenues to cut preexisting taxes. This double offset helps ensure that auctioned permits can generate an overall efficiency improvement.

This issue deserves further close study, given the prominence of emissions trading in the discussions of future international and national efforts to reduce carbon emissions.

Technological Change

There has been a good deal of discussion about how the potential for induced technological change might substantially lower, and perhaps even eliminate, the costs of CO_2 abatement policies. These discussions have exposed very divergent views as to whether technological change can be induced at no cost, or whether a resource cost is involved.

Goulder and Schneider (1996) examine these issues in a dynamic general-equilibrium framework. Knowledge capital enters the production function of each industry much like ordinary physical capital and other inputs (labor, energy, and materials). The model considers the supply of knowledge-generating resources (skilled engineers, analysts, and consultants) as well as the demand. Since such resources are scarce, there is a resource cost involved in increasing the aggregate supply of knowledge-generating resources. Similarly, at any moment in time, if one industry bids R&D resources away from another, the acceleration in technological progress in the expanding industry is offset to some degree by the slow-down in technological progress in the other.

In this model, the presence of *induced technological change* (ITC) generally lowers the costs of achieving a given abatement target. At the same time, the gross costs of a given carbon tax are generally (see qualification below) higher in the presence of induced technological change than in its absence. In the presence of induced technological change, the economy responds more "elastically" to the carbon tax and endures greater costs in response to the tax. Although this heightened elasticity implies larger gross costs to the economy, it also implies larger net benefits, because the more elastic adjustment implies greater carbon abatement than would occur in the absence of ITC.

Under special circumstances ITC can reduce the gross costs of a carbon tax. Specifically, if R&D has been substantially weighted toward the fossil fuel industries before the imposition of a carbon tax, then a carbon tax can reduce this allocative inefficiency. As a result, the costs of a carbon tax can be quite low or even negative. A substantial prior misallocation toward carbon-intensive industries could occur if there were prior subsidies toward R&D in the fossil fuel industries (with no comparable subsidies in other industries), or if there were substantial positive spillovers from R&D in noncarbon industries (with no comparable spillovers in the fossil fuel industries). Under plausible initial conditions, however, induced technological change raises, rather than lowers, the costs of any given carbon tax.

The same model has been employed to compare the costs of achieving a given abatement target through carbon taxes and R&D subsidies (Goulder and Schneider 1996; Schneider and Goulder 1997). If there are no spillovers to R&D, the least-cost way to reach a given abatement target is through a carbon tax alone. The carbon tax best targets the climate-change-related externality from combustion of fossil fuels, and thus is most cost effective. On the other hand, if there are spillovers to R&D, the least-cost way to achieve a given abatement target is through the combination of a carbon tax and an R&D subsidy. If spillovers are present, there is a market failure in the R&D market as well as a (climate-change-related) market failure associated with the use of carbon. Two instruments—the R&D subsidy and the carbon tax—are needed to address the two distinct market failures most efficiently. In general, an R&D subsidy by itself will not offer the least-cost approach to reducing carbon emissions. Results from this model are highly sensitive to assumptions about the nature and extent of knowledge spillovers. Further empirical work that sheds light on these spillovers would have considerable value.

Recent papers by Nordhaus (1997) and Goulder and Mathai (1997) examine how the presence of induced technological change affects the optimal carbon tax profile. Both papers indicate that the possibility of induced technological change has a very minor impact on the optimal tax profile. This reflects the fact that such change has only a small impact on the time profile of CO_2 *concentrations* relative to the time profile of such concentrations that would occur in the absence of ITC. Thus, induced technological change has only a small impact on the marginal damage from such concentrations at each point in time. Since the optimal carbon tax at each point in time equals the marginal damage, the optimal tax profile is not much affected. Nordhaus finds, further, that induced technological change has an extremely small impact on levels of abatement and the net gains from carbon taxes at the optimum, whereas Goulder and Mathai find that these effects are sensitive to the specification of abatement cost functions. Further research to investigate more closely the underlying empirical issues might yield significant policy insights.

The returns to R&D are very uncertain, but most empirical studies suggest underinvestment in R&D (see, for example, Sakurai and others 1996). This underinvestment probably results from the private sector's assessment that they could not appropriate many of the benefits of the R&D, and from the government's and public's lack of belief or understanding of this rationale for government R&D support. Direct public support for R&D would involve large direct costs to the government but would avoid the indirect market costs (deadweight costs) resulting from energy/carbon taxation. On the other hand, public R&D support might simply displace private R&D support. Moreover, the government's track record for selecting large-scale R&D technologies to fund has not generally been very good, which suggests the government should fund more small-scale demonstration projects and let the private sector determine which of the successful ones to fund at the next stage.

Capital Availability and Flows

Not much thought has gone into the economywide drain on capital (or skilled labor) that would result from the large-scale emissions-reduction programs being contemplated. What look like good investments at the micro level may have less appeal at the macro level, where capital may be in limited supply in some key countries. The current costs of capital in different countries reflect the extent to which these supply constraints operate. But it is difficult to project how these constraints will evolve in the future, as they depend partly on market forces and partly on policy choices made by the home country and its trading partners.

Macroeconomic Impacts of Carbon Emissions Reduction

Most analyses of the timing issue have been done with full-employment models, which Jaccard and Montgomery (1996) report may significantly understate the short-run costs of abruptly imposing carbon emissions limits. According to results from macroeconometric simulations reported in their paper, the cost of immediately stabilizing emissions might be several times higher when the unemployment created by subjecting the macroeconomic system to sudden shocks is taken into account (see also Rose and Liu 1995). The additional unemployment results from the propensity of wages in modern economies to resist decreases that occur in response to reductions in the value of labor inputs (often referred to as inflexible wages) caused by external shocks like higher energy prices or taxes. Being unable to reduce the price of labor, firms then reduce the quantity demanded in order to increase the value of the remaining workers relative to the prices of other inputs. This effect can be offset to some extent if the Federal Reserve Board "accommodates" the shock by increasing the money supply. In effect, this response increases inflation, thereby reducing real wages and prices. On the other hand, increasing inflation can negatively impact the economy as well—affecting consumption and investment behavior—and inflationary impulses are often followed by monetary tightening.

The Federal Reserve Board can, therefore, play a major role in determining the economic impacts of carbon emissions control programs that work through changes in energy prices. These impacts need to be studied in more detail before policymakers will commit to a carbon emissions reduction program that goes very far beyond no-regrets measures.

Economic Growth Projections

Most projections of economic growth seem disconnected from either historical performance or systematic projections based on growth accounting/endogenous growth theory, and so on. A rapid convergence in economic performance among nations is often assumed. We need to ask ourselves why countries in transition are in transition and why developing countries are developing countries (and what has changed). This will unavoidably lead to good economic prospects in some countries and not-so-good prospects in others but, in general, more realistic assessments of where and when growth in carbon emissions will occur and who could and should reduce them. The international trade and competitiveness implications of carbon emissions reduction policies are related issues that need additional attention (see, for example, Manne and Rutherford 1992; McKibben and Wilcoxen 1992; Nicoletti and Oliveira-Martins 1992).

Fossil Fuel Resource Base Estimates

The size and distribution of the global resource base for fossil fuels comprise another set of key inputs to the projection process concerning the carbon emissions/cost of carbon emissions mitigation. In the short run, large amounts of low-cost accessible gas could result in relatively low-cost options to reduce emissions by switching from coal (and to some extent oil) to natural gas. In the longer term, the extent of accessible and economic coal reserves is critical to projections of the maximum cumulative long-term carbon emissions levels in the absence of mitigation policies. Estimates of the ultimate worldwide oil and gas resource base are highly uncertain, especially outside the OECD, and these uncertainties are compounded by differences in definitions between geologists and economists. The uncertainties regarding the extent of the global coal resource base, which is widely believed to be as much as an order of magnitude larger than that for oil or gas, is even more uncertain.

Population Growth Projections

Population growth projections are also critical to emissions projections and emissions-reduction cost projections. Population growth relates to other societal goals (especially in developing countries) and acts as a key climate-change policy lever (as the commitments many developing countries made at world population summits over the last few years demonstrate). Unfortunately, most climate-change policy analyses are based on a few official forecasts published by the United Nations and the World Bank. We need to learn more about how these projections were developed and why, so alternative paradigms can be developed and tested.

Future population levels are fundamental to the analysis of long-run resource and environmental problems. Therefore, in addition to better exogenous projections of population growth, we need to include in our analyses ways to represent the likely impacts of climate change on population growth. Even more importantly, it would be extremely useful to develop options for using population control to counter global warming and other long-run global problems.

Discount Rates

The choice of appropriate discount rates is addressed elsewhere in the WGIII report (Arrow and others 1996), but it is also an extremely important parameter in the analysis and design of appropriate climate-change policies. Since the costs of mitigation start almost immediately and most of the benefits are projected to show up only after a number of decades, the discount rate becomes a critical parameter in making decisions on the appropriate amount of market intervention. Given the residual uncertainty about this parameter and its schizophrenic status as both an external parameter and a policy instrument, more uncertainty analysis on a single rate and more experimentation with various alternative implementations would be quite valuable. For example, some research has proposed that direct assessments of people's willingness to pay for climate-change policies across time and space be employed (for example, Schelling). Others (for example, Nordhaus) have suggested that any move from the current descriptive rate (of 5% to 6%) to a much lower normative rate should take place quite gradually over a number of decades, rather than immediately, because the existing capital stock and mix of inputs to the economy are based on the higher rate and will last for many years.

SUMMARY

The research summarized in Chapter 9 of the Working Group III contribution to the IPCC's second assessment report suggests several important conclusions regarding the likely costs of future efforts to reduce emissions of greenhouse gases.

First, a least-cost set of policies and measures designed to stabilize or gradually reduce global emissions of greenhouse gases would probably cost only about 1% of global economic output. Second, the least-cost emissions-reduction program needs to be flexible as to where, when, and how the emissions reductions are implemented because costs vary and are uncertain in all three dimensions. A rigid program specifying in advance exactly when, where, and how to undertake emissions reductions could easily cost four, five, or more times the least-cost program. Finally, additional research on technology invention, innovation, and diffusion could improve our understanding of the costs of reducing greenhouse gas emissions and of how to design a least-cost emissions-reduction program.

REFERENCES

Arrow, K. J., W. R. Cline, K.-G. Mäler, M. Munasinghe, R. Squitieri, and J. E. Stiglitz. 1996. Intertemporal Equity, Discounting, and Economic Efficiency. In IPCC 1996.

Bovenberg, A. Lans and Lawrence H. Goulder. 1996. Optimal Environmental Taxation in the Presence of Other Taxes: General Equilibrium Analysis. *American Economic Review* 86(4): 985–1000.

———. 1997. Costs of Environmentally Motivated Taxes in the Presence of Other Taxes: General Equilibrium Analysis. *National Tax Journal* 50(1).

Bovenberg, A. Lans and Ruud de Mooij. 1994. Environmental Levies and Distortionary Taxation. *American Economic Review* 84(4): 1085–89.

EPRI (Electric Power Research Institute). 1990. *Residential Customer Preference and Behavior: Market Segmentation Using CLASSIFY*. Rep. EM-5908. Palo Alto, California: EPRI.

Farhar-Pilgrim, B. and C. T. Unseld. 1982. *America's Solar Potential: A National Consumer Study*. New York: Praeger.

Goulder, Lawrence H. 1995. Environmental Taxation and the Double Dividend: A Reader's Guide. *International Tax and Public Finance* 2: 157–83.

———. 1996. The Goulder Model: Structure and Simulations Results. In *Reducing Global Carbon Emissions: Costs and Policy Options*, edited by Darius Gaskins and John Weyant. Stanford, California: Energy Modeling Forum, Stanford University.

Goulder, Lawrence H. and Koshy Mathai. 1997. Optimal CO_2 Abatement in the Presence of Induced Technological Change. Department of Economics, Stanford University, October.

Goulder, Lawrence H., Ian Parry, and Dallas Burtraw. 1997. Revenue Raising vs. Other Approaches to Environmental Protection: The Critical Significance of Pre-Existing Tax Distortions. *RAND Journal of Economics* 28(4): 708–31.

Goulder, Lawrence H. and Stephen H. Schneider. 1996. *Induced Technological Change, Crowding Out, and the Attractiveness of CO_2 Emissions Abatement.* Institute for International Studies, Stanford University, Stanford, California, October.

Huntington, Hillard, Lee Schipper, and Alan H. Sanstad. 1994. *Markets for Energy Efficiency.* Special issue of *Energy Policy* 22(10).

IPCC (Intergovernmental Panel on Climate Change). 1996. *Climate Change 1995: Economic and Social Dimensions of Climate Change.* Edited by James P. Bruce, Hoesung Lee, and Erik F. Haites. Cambridge: Cambridge University Press.

Jaccard, M. and D. W. Montgomery. 1996. Mitigation Costs for the United States and Canada. *Energy Policy* 24(10/11): 889–98. (In special issue, *Energy and Greenhouse Gas Mitigation: The IPCC Report and Beyond,* edited by Erik F. Haites and Adam Rose.)

Jaffee, Adam B. and Robert N. Stavins. 1994. The Energy Efficiency Gap: What Does It Mean? In Huntington, Schipper, and Sanstad 1994.

Jorgenson, Dale W. and Peter J. Wilcoxen. 1995. Reducing U.S. Carbon Emissions: An Econometric General Equilibrium Assessment. In *Reducing Global Carbon Emissions: Costs and Policy Options,* edited by Darius Gaskins and John Weyant. Stanford, California: Energy Modeling Forum, Stanford University.

Kempton, W. 1992. Social Science Research Recommendations: Buildings Energy Efficiency Program Review. Newark: University of Delaware.

———. 1993. Will Public Environmental Concern Lead to Action on Global Warming? *Annual Review of Energy and the Environment* 18:217–43.

Kempton, W. and L. L. Lane. 1994. The Consumer's Energy Analysis Environment. *Energy Policy* 10:857–66.

Kempton, W. and L. Montgomery. 1982. Folk Quantification of Energy. *Energy: The International Journal* 7(10): 817–27.

Leonard-Barton, D. 1981. The Diffusion of Active Residential Solar Energy Equipment in California. In *Marketing Solar Energy Innovations,* edited by A. Shama. New York: Praeger.

Leonard-Barton, D. and E. M. Rogers. 1981. *Horizontal Diffusion of Innovations: An Alternative Paradigm to the Classical Diffusion Model.* Working Paper 1214–81. Cambridge, Massachusetts: Sloan School of Management, MIT.

Lutzenhiser, L. 1992. A Cultural Model of Household Energy Consumption. *Energy: The International Journal* 17:47–60.

———. 1993. Social and Behavioral Aspects of Energy Use. *Annual Review of Energy and the Environment* 18:247–89.

Manne, A. S. and Richard G. Richels. 1992. *Buying Greenhouse Insurance.* Cambridge, Massachusetts: MIT Press.

Manne, A. S. and T. F. Rutherford. 1992. *International Trade in Oil, Gas and Carbon Emission Rights: An Intertemporal Equilibrium Model.* Paper presented at the International Workshop on Costs, Impacts and Possible Benefits of CO_2 Mitigation. September, Vienna, International Institute for Applied Systems Analysis.

McFadden, Daniel. 1990. Residential Customer Preference and Behavior: Market Segmentation Using CLASSIFY. Rep-EM-5908. Palo Alto: Electric Power Research Institute.

McKibbin, W. J. and P. J. Wilcoxen. 1992. *G-Cubed: A Dynamic Multi-Sector General Equilibrium Model of the Global Economy.* Discussion Paper in International Economics, No. 98. November. Washington, D.C.: The Brookings Institution.

Metcalf, G. E. 1994. Economics and Rational Conservation Policy. *Energy Policy* 22(10).

Mongia, P. and others. 1991. Cost of Reducing CO_2 Emissions from India. *Energy Policy* 19(10): 978–87.

Nicoletti, G. and J. Oliveira-Martins. 1992. *Global Effects of the European Carbon Tax.* Technical Report No. 125. OECD Department of Economics and Statistics, Resource Allocation Division, Paris.

Nordhaus, William D. 1993. Optimal Greenhouse Gas Reductions and Tax Policy in the "Dice" Model. *American Economic Review* (May).

———. 1994. *Managing the Global Commons: The Economics of Climate Change.* Cambridge, Massachusetts: MIT Press.

———. 1997. Modeling Induced Innovation in Climate-Change Policy. Paper presented at the IIASA/NBER Workshop on Induced Technological Change and the Environment, International Institute for Applied Systems Analysis, Laxenburg, Austria, June 26–27.

Parry, Ian W. H. 1995. Pollution Taxes and Revenue Recycling. *Journal of Environmental Economics and Management* 29(3): 564–77.

Parry, Ian W. H., Roberton C. Williams, and Lawrence H. Goulder. 1996. When Can Carbon Abatement Policies Increase Welfare? The Fundamental Role of Distorted Factor Markets. NBER working paper, no. 5967. Cambridge, Massachusetts: National Bureau of Economic Research.

Robinson, J. B. 1992. Of Maps and Territories: The Use and Abuse of Socioeconomic Modeling in Support of Decision Making. *Technological Forecasting and Social Change* 42:147–64.

Robinson, J. B. and P. Timmerman. 1991. Myths, Rules, Artifacts, EcoSystems: Framing the Human Dimensions of Global Change. Paper presented at the conference on Human Responsibility and Global Change, Goteborg, Sweden, June 9–14.

Rogers, E. M. 1963. *The Diffusion of Innovation.* New York: Free Press.

Rose, Adam and S. Liu. 1995. Regrets or No Regrets—That Is the Question: Is Conservation Really a Costless CO_2 Mitigation Strategy? *Energy Journal* 16:67–87.

Sakurai, Norihisa, George Papaconstantinou, and Evengelos Ioannidis. 1996. The Impact of Technology Diffusion on Productivity Growth: Empirical Evidence for 10 OECD Countries in the 1970s and 1980s. STI Working Paper 1996/2. Directorate for Science, Technology, and Industry, Organisation for Economic Cooperation and Development, Paris.

Schneider, Stephen H. and Lawrence H. Goulder. 1997. Achieving Carbon Dioxide Concentration Targets: What Needs to Be Done Now? *Nature* 389 (September 4): 3–4.

Shackleton, R., M. Shelby, A. Cristofaro, R. Brinner, J. Yanchar, L. Goulder, D. Jorgenson, P. Wilcoxen, P. Pauly, and R. Kaufmann. 1996. The Efficiency Value of Carbon Tax Revenues. In *Reducing Global Carbon Emissions: Costs and Policy Options*, edited by Darius Gaskins and John Weyant. Stanford, California: Energy Modeling Forum, Stanford University.

Stern, P. C. 1986. Blind Spots in Policy Analysis: What Economics Doesn't Say About Energy Use. *Journal of Policy Analysis and Management* 5:200–227.

Stern, P. C., E. Aronson, J. M. Darley, D. H. Hill, E. Hirst, and others. 1986. The Effectiveness of Incentives for Residential Energy Conservation. *Evaluation Review* 10:147–76.

Stern, P. C. and S. Oskamp. 1987. Managing Scarce Environmental Resources. In *Handbook of Environmental Psychology*, vol. 2, 1043–88. New York: Wiley.

U.S. DOE (Department of Energy). 1992. *Energy Conservation Trends: Understanding the Factors that Affect Conservation Gains in the United States Economy.* DOE/PE–0092. Office of Conservation and Renewable Energy. Washington, D.C.: DOE.

Weyant, J. 1998. Energy and Industrial Systems. In *Human Choice and Climate Change: An International Assessment.* Pacific Northwest National Laboratory.

Weyant, J. and Y. Yanigisawa. 1996. Energy and Industry. Chapter 4 in S. Rayner and E. L. Malone (eds.) *Human Choices and Climate Change: Volume 2, Resources and Technology.* Columbus, Ohio: Batelle Press.

Comments

The Costs of
Greenhouse-Gas Abatement and
Carbon Emissions Reductions

Richard Richels

In most areas, I am in agreement with Professors Kram and Weyant. This is particularly the case with regard to the limitations of the "top-down/bottom-up" discussion. In retrospect, the decision to separate the methodology discussion (Chapter 8 of WGIII) from the reporting of numerical results (Chapter 9 of WGIII) may have been a mistake. In Chapter 8, we explain the key differences between the two approaches, describe their relative strengths and weaknesses, and suggest how the methodologies might be constructively combined to maximize their contribution to the debate. However, we fail to take advantage of the insights contained in the Chapter 8 discussion when presenting specific numerical results in Chapter 9.

One way to bridge the gap between the two chapters would have been to make better use of the so-called Kaya identity, which states that the growth rate in CO_2 emissions is equal to the growth rate in GDP less the decline rate in energy per unit of economic output less the decline rate in CO_2 per unit of energy. Chapter 9 uses the Kaya identity to summarize the results of the EMF-12 models. The analysis is for the United States and covers the period 1990–2010.

In the EMF-12 reference case, carbon emissions and GDP grow at average annual rates of 1.0% and 2.2%, respectively. In the absence of a

RICHARD RICHELS is Director of Global Climate Change Research at the Electric Power Research Institute.

carbon constraint, there is some minor fuel switching toward more-carbon-intensive fuels but this is more than offset by the decline in energy use per unit of economic activity. In the stabilization case, the constraint induces sufficient investment in supply-side substitutes and conservation to eliminate growth in carbon emissions. This is done without significantly reducing the growth of the economy. Whereas the losses due to the constraint are large in absolute terms, they are relatively small when measured in terms of a reduction in the GDP growth rate.

The Kaya identity can be used to explain why the top-down and bottom-up results differ so markedly. Bottom-up modelers are either more optimistic about the price of conservation or more optimistic about the cost and availability of less-carbon-intensive supply-side options, or both. The reason some bottom-up studies show economic gains from a carbon constraint is that they assume that supply- and demand-side breakthroughs will occur in the policy case but not in the reference case.

The two approaches, however, are not mutually exclusive. The more engineering-oriented approach is particularly helpful in sharpening our understanding of the market potential of emerging technologies—on both the supply and demand sides of the energy sector. These results provide valuable inputs into top-down analyses, which enhance our understanding of the economywide impacts of a carbon constraint.

This brings us to the issue of hybrid models. We were not sufficiently explicit about the extent to which such models already exist and are currently being engaged in the policy debate. A good example is provided by those general equilibrium models that attempt to provide a bottom-up perspective of the supply side of the energy sector by describing energy technologies in process model detail. For the balance of the economy, these models tend to take a more top-down perspective. Here, bottom-up models could be particularly helpful in providing insights into the technical progress component of the so-called AEEI (autonomous energy efficient improvements).

I also agree with the reviewers' comments that we insufficiently stressed the extent to which existing studies may underestimate costs. For example, Weyant notes that by relying on general equilibrium models for calculating the value of flexibility in timing, we may have underestimated the benefits. This is because we ignore the short-term shocks of a near-term carbon constraint. Unfortunately, the literature did not include such an analysis at the time of the study—at least, as applied to the timing issue. We should have nevertheless acknowledged the problem.

Kram notes that Chapter 8 implies that "double dividends" are exclusively positive. He points out that inefficient tax-recycling schemes or inappropriate redistribution of R&D could have negative effects. Although the possibility of the former is discussed in Chapter 9, more attention should have been devoted to this topic.

I agree with the reviewers that the issue of endogenous technical change deserved more attention—albeit, at the time of the Second Assessment Report (SAR), the literature was spotty on this topic as it relates to climate change. Nevertheless, we could have done a better job of framing the issue. At the very least, we could have distinguished between two schools of thought. On the one hand, there are those who believe that a tax or near-term constraint will convince the private sector of the need to reduce emissions and hence encourage the development of low-cost alternatives sooner than might have otherwise been the case. People belonging to this school tend to underemphasize the appropriability issue and the deadweight losses induced by the constraint. On the other hand, there are those that have more faith in the government than in the private sector to shepherd technology through the R&D process. They tend to underemphasize negative aspects of the government's track record to date.

Kram is correct that the focus on CO_2 is justified because of its overwhelming importance as a greenhouse gas. However, from the perspective of conducting comprehensive integrated assessments, it is imperative that other gases and sink enhancement be given more attention as mitigation options. Otherwise, we may be ignoring inexpensive sources for greenhouse-gas reductions. We were well aware of this shortcoming but were frustrated by the paucity of research in this area.

Although I am in general agreement with the reviewers in most instances, there are several areas in which we differ. First, I do not find the IPCC format, as presently constituted, conducive to resolving issues such as the size of the "no regrets" option. This is not a question of literature review. It is an issue requiring substantial new research. Second, Kram raises the question of minimum requirements for models to adequately address particular issues. The suggestion is that Chapter 9 should include some consideration of the appropriateness of a particular model to conduct a given study. I believe that we did do this to some extent in our choice of studies for review. Going much further than this was beyond the scope of the exercise. Third, in light of the scores of studies reviewed, it would have been impractical to critique the choice of input assumptions for each modeling team. We did, however, try to provide insight into how alternative assumptions would affect a study's results.

In summary, Professors Kram and Weyant offer a number of constructive and insightful criticisms regarding Chapters 8 and 9 of the WGIII report. Indeed, their suggestions would have been very useful at the time of the exercise. This raises the issue of the timing and nature of the review process. The IPCC should give careful consideration to formalizing a Snowmass-like process and scheduling it at an earlier stage in the assessment process.

8

Climate-Change Damages

Robert Mendelsohn

INTRODUCTION

Determining the optimal strategy for controlling greenhouse gases requires weighing the costs of abating these pollutants against the damages that the pollutants would otherwise cause. Although reasonably accurate measures of abatement are available for reducing greenhouse-gas emissions over time—see Tom Kram's paper in this volume on abatement costs—measuring the damages from greenhouse-gas emissions is more elusive. Damage estimates require integrating climate science, ecological responses, human reactions, and social values. The process involves long lags separating emissions from final consequences. The magnitude of the effects will vary across space, improving some areas and making other areas worse off. The impacts affect many sectors of the economy and many aspects of the quality of life. The impact process is dynamic, implying impacts will change over time. The impacts will affect people from many different cultures with widely varying incomes and circumstances. The values to be applied to changes will therefore vary by location, and people will disagree about which values to choose. Not surprisingly, the damages from climate change are controversial. Yet, it is imperative that the world be convinced that substantial benefits arise from controlling greenhouse gases, because there are many political and economic barriers to creating effective control programs.

The IPCC, in its recent assessment of the state-of-the-science surrounding climate change, provides two perspectives on climate-change impacts. First, Working Group II (WGII) of the IPCC provides detailed accounts of how climate change might affect individual sectors of the

---

ROBERT MENDELSOHN is the Edwin Weyerhaeuser Davis Professor at the Yale School of Forestry and Environmental Studies.

world economy and various aspects of the global ecology (Watson and others 1996). Second, Working Group III (WGIII) estimates the social damages associated with these physical changes (Bruce, Lee, and Haites 1996, in particular Chapter 6). Although the individual impact studies provide a wealth of information about the detailed reaction of different ecological and economic systems to climate change, the sheer magnitude of the information about all the different sectors for every country in the world is overwhelming. These complex physical impacts need to be understood and compared against the cost of various abatement programs for policymakers to design effective programs. Although it is important to collect the evidence of the effects of climate change, a mere listing of these consequences is cumbersome and overwhelming. The size of the WGII report on physical impacts is proof enough that the complexity of global climate-change impacts is simply too massive to present in a raw form. No policymaker could read this report and deduce which abatement program makes sense for his country, much less the world. In order to use the complex scientific impact results in policy, analysts must convert these myriad physical impacts into a useful damage estimate. The authors of Chapter 6 of the WGIII report accepted this formidable challenge and confronted the controversial but necessary step of valuing climate-change impacts across the globe.

With its strong natural science foundation, the climate-science community is understandably reluctant to commit itself to placing values on environmental effects. However, in order to decide whether climate change is harmful to society and whether dramatic mitigation efforts are worthwhile, one cannot escape from applying values to the set of physical changes that climate will induce. Applying values to damage estimates thus represents a controversial but necessary activity. Given the relatively high cost assigned to abatement control, the benefits of control programs must be substantial to justify their expense. Damage studies identify what these benefits could be and quantify their magnitude. The process organizes the vast impact material and highlights important impacts, focusing scientists and policymakers alike on the underpinnings and evidence of the most critical effects. By explicitly stating values, damage studies also invite criticism, but these debates are at the heart of environmental policy. It is far better to openly debate values and subject them to the scrutiny of evidence and logic than to refuse to acknowledge values and yet allow policy to be driven by them.

In this paper, I review the IPCC estimates of climate-change damages. The next section summarizes the methods of the IPCC and reviews its findings. After that, in the section Critique of Conceptual Approach, I evaluate the strengths and weaknesses of the analytical approach. In the fourth section, I critically review the empirical methodology of the report.

Taking more recent scientific and economic research into account, I offer an alternative view of impacts in the final section.

IPCC DAMAGE ESTIMATES

Economic logic implies society should minimize the present value of the damages and abatement costs from all pollution activities. This is just a commonsense notion that society should try to keep the total costs associated with pollution as low as possible. However, in order to implement this idea, one must be careful to measure all costs. Abatement costs, because they entail modifying market activities, are relatively easy to measure. Damage costs, because they involve complex environmental links and often entail the loss of nonmarket as well as market services, are more difficult to measure. An incomplete accounting of damages would bias decisions made according to this calculus because damages would be underestimated. The pollution-control literature is consequently quite concerned with the monetization of damages. Climate change, because of its enormous spatial scale, complexity, and intertemporal qualities, represents the most challenging impact analysis ever undertaken.

Chapter 6 of the IPCC WGIII report, Social Costs of Climate Change, monetizes the damages expected to occur from climate change. Although the authors briefly review alternative approaches to choosing abatement policies, they argue that cost-benefit analysis provides important insights which help guide policymakers to more efficient decisions. Cost-benefit analysis requires that both the costs and the benefits of pollution control be reduced to a common index. Since the abatement costs have already been quantified in dollars, it makes sense to quantify the benefits in dollars as well. All the significant benefits must be quantified. This requires valuing both market and nonmarket impacts. Market impacts affect sectors of the economy such as agriculture and forestry, whereas nonmarket impacts concern quality-of-life issues such as health, species loss, and amenities. These arguments are both logical and quite consistent with economic theory.

Although the authors acknowledge that climate control is a dynamic phenomenon, they rely heavily on a comparative static equilibrium approach in their analysis. Most of the analysis is limited to a comparison of two steady states: welfare under current climate conditions versus welfare under an equilibrium climate associated with the doubling of greenhouse gases ($2 \times CO_2$). These two steady states often assume current economic conditions. That is, they do not forecast what the economy will look like in the future when climate changes are predicted to take place. Since climate is expected to change gradually over the next century, there are serious problems with using the current economy as the baseline (control scenario). However, this flawed approach is the norm in most of the impacts literature, so one cannot criticize the IPCC too severely for perpetuating this shortcoming. However, they could have done a better job of explaining the fundamental difference between this comparative static approach and the dynamic optimal control approach, which they also review.

The authors review the many sectoral studies in agriculture and sea-level rise and the few studies which have been done on forests, water supply, energy, and other market effects. They also briefly review each of the nonmarket impacts including health, air pollution, water pollution, migration, human amenities, and ecosystem and biodiversity impacts. Perhaps because so much more research has been done on farming and sea-level rise, the chapter spends most of its critical energies on these two sectors. Other sectors, such as biodiversity and migration, are supported by almost pure speculation, perhaps because so little can reliably be said about their values. In any case, the nonmarket impacts get a light touch in the chapter.

The chapter also reviews selected regional studies of impacts from around the world. These studies are comparative static analyses, examining current and $2 \times CO_2$ climates. However, the bulk of the regional studies are based on older climate-change scenarios that assume temperature changes closer to $4.5°C$ with reductions in summer precipitation, runoff, and sea-level rise of one meter. Given recent advances in climate science, these older scenarios are now considered extreme forecasts (Houghton and others 1996). Further, most of the empirical studies reviewed were robot models which assumed only limited adaptation by human populations across decades. Chapter 6 makes no attempt to adjust the results of these studies to relate to a common climate scenario or preferred approach.

Further, Chapter 6 does not appear to rely on any of the cited empirical studies directly. Rather, the authors appear to depend almost exclusively on several comprehensive judgments (Nordhaus 1991; Cline 1992; Fankhauser 1995; Titus 1992; and Tol 1995) that have been made on the basis of earlier literature. These estimates for the United States are reported in the chapter's Table 6.4. The U.S. damages range from $55 billion to $139 billion ($1990) from the doubling of greenhouse gases. Except for the estimate by Titus (1992), the different authors arrive at similar totals for the United States, arguing that total damages range from 1.1% to 1.5% of GDP. Although the total estimates agree, there is no consistency across the authors about the magnitude of individual sectors. The authors estimate that market damages (agriculture, forestry, sea-level rise, energy, and water availability) amount to 25% to 73% of the total damages from climate change.

Unfortunately, no country other than the United States has been the subject of a comprehensive set of empirical impact studies of each sector. Nonetheless, Tol (1995) and Fankhauser (1995) have taken the U.S. results, scattered regional studies, and some regional data and have extrapolated from the United States to the rest of the world (in $1990). These two authors estimate that annual OECD damages would be about $180 billion to $190 billion (1.3% to 1.6% of world GDP) from $2 \times CO_2$. Estimates of impacts to Eastern Europe and the rest of the former Soviet Union range from –$8 billion to $18 billion (–0.3% to 0.7% of GDP). Estimates of impacts to the rest of the world range from $71 billion to $134 billion (2.4% to 6.6% of GDP). The total damages for the entire world from doubling are estimated to be between $270 billion and $316 billion (1.4% to 1.9% of GDP).

Chapter 6 also discusses what value to place on a ton of carbon emissions. The authors report the results of seven studies. Five of the estimates (Nordhaus 1991, 1994; Peck and Teisberg 1992; Fankhauser 1995; and Maddison 1994) rely on optimal-control models. The optimal-control estimates explicitly recognize the dynamic nature of this problem and seek to determine the minimum total cost solution. Consequently, the optimal-control models carefully evaluate the damage done by a ton of carbon emissions over time. These dynamic studies explicitly account for the fact that greenhouse gases are slowly accumulating and remain in the atmosphere for many decades. A ton of emissions thus causes a stream of damages. This stream must be evaluated relative to the marginal cost of abatement at the moment of emission. The optimal-control models take these dynamic features into account. The models yield expected values from $5 to $20 per ton for the next decade and from $7 to $27 per ton in three decades. The two alternative estimates reported in Chapter 6 suggest values from $30 to $125 per ton. These estimates are higher because they assume very low discount rates and that carbon emissions will increase rapidly for centuries.

Chapter 6, on social costs, argues that many secondary benefits would result from control policies because they would reduce fossil fuel consumption and thus reduce other emissions as well. Specifically, the authors argue that reductions in sulfur dioxide, particulates, carbon monoxide, and nitrogen oxides could yield a range of secondary benefits between $2 to $500 per ton of carbon depending on the program and country.

The authors are careful to acknowledge that many uncertainties surround impact estimates, including regional variation, climate sensitivity, absence of empirical research, long-term consequences, and future developments. They also stress the possible role of catastrophes and other climatic and ecological surprises.

The authors do not present a complete description of how they move from their review of the empirical studies to their bottom-line estimates. The chapter appears to assume that two of the authors (Tol 1995 and Fankhauser 1995) have done a good job of making these judgments and simply reports their results. A doubling of CO_2 in the atmosphere is predicted to cause global damages of 1.5% to 2.0% of world GDP. These impacts would fall more heavily on developing nations (2% to 9% of GDP) than developed countries (1% to 1.5% of GDP). They caution that further increases in CO_2 beyond doubling could lead to climate increases of 10°C, which would reduce world GDP by 6% or more. The authors argue that the damage per ton of carbon emissions ranges from $5 to $125, although at a 5% discount rate, the value is limited to $5 to $12 per ton. Finally, the authors argue that the secondary benefits of removing other fossil fuel pollutants add $2 to $500 of benefits per ton of carbon removed.

CRITIQUE OF CONCEPTUAL APPROACH

Although any attempt to value (monetize) the environmental damages from global warming will meet intense criticism, the authors should not be criticized simply because they were willing to undertake this formidable task. Greenhouse-gas concentrations are largely a public good. No single individual or country has control over the global emissions of greenhouse gases. Global emissions are the outcome of the actions of nations across the earth, and the entire earth must share in the consequence. As with all public goods, people disagree about what weight to give these consequences. Climate change is not unique in representing a public good. But the huge dimensions of climate change, affecting every country in the world and many generations, exacerbate some of these problems. Even within many nations, intense controversies surround environmental values. Across countries with such different cultural backgrounds and incomes, the range of held values are certain to be wider and their resolution even more emotional.

Choosing the desired level of public goods is further exacerbated by incentives for countries to act strategically to gain some advantage. For example, countries and individuals who suffer a disproportionate share of the impacts or who are particularly sensitive to the impacts may exaggerate their estimate of the damages. If they can exaggerate the magnitude of the damages, it will warrant large abatement programs. Nations, people, and industries who would have to pay a disproportionate share of potential abatement costs, however, may argue these impacts are small and unimportant and warrant only minuscule abatement programs designed more for public relations than effectiveness. Because huge stakes are involved, nations have incentives to act strategically, making the determination of damages a highly political action.

If people disagree about what values to place on environmental impacts, what rational methodology is available to assess a public good such as global warming? From the seminal paper by Samuelson (1964), we gain the simple but critical insight that the social damage from a public good is the sum of the individual damages assessed by all affected people. People do not need to be forced to agree on a single value; the social calculus can allow for a broad range of values held by the population. The social damage function is the *sum* of the individual damage functions.

The correct theoretical approach is to measure the values of the people affected by each impact in every country and sum these values across all people. The result will be an aggregate valuation of global impacts. Ideally, damages in every country could be measured and then summed. The absence of such damage estimates forces the authors to take values measured in the United States and extrapolate them to the rest of the world. Although this might appear to be an example of American values being thrust upon the world, the authors try to adjust these estimates to each country. For example, many of the market effects are calibrated to the size of the sector in each economy. Nonmarket impacts are assumed to vary with per capita income. Although one could criticize the accuracy of these extrapolations, the basic concept that the impacts should be valued by the people affected is a sound analytical approach.

The most controversial application of this approach is probably the valuation of human life. Several recent scientific studies suggest that global warming has the potential to increase human mortality and morbidity rates (see Watson and others 1996). Countries face the prospect of preventing premature deaths all the time in their health, welfare, and pollution-control policies. The authors argue that the values countries exhibit in these decisions reflect what value to associate with global warming impacts. Poor countries, with limited resources and many competing claims, can and do spend only small amounts protecting life whereas rich countries can afford to spend vast amounts on small increases in life expectancy. The authors embody this discrepancy in values in their calculations. Some critics have argued that it is immoral to place different values on protecting lives in different countries. However, this unequal outcome is a fair choice if incomes remain unequal (Esty and Mendelsohn 1998). Because poor countries cannot afford to spend vast sums of money on small improvements in life expectancy, to force them to do so would make them distinctly worse off. Further, to spend vast sums to protect poor people from one threat while spending little to protect them from far more serious health risks which happen to be domestic and not international is both inefficient and unfair. Although it is reasonable to be concerned about equalizing economic well-being across people, arbitrarily assuming equal values for small health risks is an ineffective mechanism to help poor people. The authors' decision to value life and other impacts using the values of the people affected is a morally defensible and economically sound approach.

The authors value both market and nonmarket impacts. Market impacts are easily monetized in that well-defined methods exist for determining the dollar value of changes in market outcomes. Economics has long studied how to reveal preferences from decisions both suppliers and consumers make as they produce and purchase goods. The fact that the changes in these patterns of behavior may be caused by climate change does not alter the soundness of these methods. For example, if climate change causes the supply of a specific agricultural crop to fall, markets will react by increasing the price until the new supply equilibrates with demand. The value of this shift can be measured by valuing the change in consumer and producer surplus, which are derived from measurements of the underlying demand and supply functions. Although questions remain about the estimation of both the underlying demand and supply functions as well as the climate-related shift in these functions, the principle is a sound economic concept.

(Economists have long debated issues surrounding compensated versus Marshallian demand functions. Although the fodder for many an academic treatise, these distinctions are rarely important empirically. For precise theoretical estimates, however, compensated demand functions must be utilized in the calculation of welfare. Another source of complexity concerns whether general-equilibrium models are required. The impacts currently being uncovered are generally not of sufficient magnitude to warrant the use of general-equilibrium models. For precise theoretical estimates, however, general-equilibrium models are preferred ; see Scheraga and others 1993).

The valuation of nonmarket goods is more difficult and controversial. Because of the tenuous nature of some of these nonmarket studies, the authors of Chapter 6 were perhaps unwilling to be critical. However, the report should probably have made it clearer which impacts were well founded and which were closer to speculation. Some nonmarket services can be valued from observed purchases of related market goods. For example, by observing the frequency of visitation as it varies with distance from a site, an analyst can deduce the demand for visiting a recreation site as though it were a market good. Alternatively, the value of small life-threatening risks can be measured by observing how much more wage earners must be paid to take slightly more risky occupations.

The authors did not, however, rely on systematic methods to value nonmarket impacts but rather utilized ad hoc approaches. Although they acknowledge that the valuation approaches are ad hoc, they do not alert the reader to just how little underlies many of these guesses. Among the most tenuous estimates are the values of "nonuse services." Nonuse services are benefits that people enjoy without any direct contact with the resource. The person holding the value has no observable behavior associated with consuming the service. The person does not buy the service, visit a specific site to enjoy the service, or spend time at an activity to obtain the service. The lack of any behavioral component makes it difficult to value. For example, the value of endangered species is difficult to measure because people take few actions that reveal how strongly they care about losing species. Social scientists have resorted to valuing nonuse, nonmarket goods with attitudinal approaches, asking people what values they place on these goods. Although these methods yield numbers, considerable debate surrounds the interpretation of attitudinal survey results for nonuse goods (see Cummings and others 1986).

One complex aspect of greenhouse gases is that they are cumulative or stock pollutants (see Nordhaus 1991; Peck and Teisberg 1992; and Falk and Mendelsohn 1993). The impact of a marginal unit of emissions depends on the stock of the pollutant in the environment—and it tends to last a long time. This leads to two important insights. First, the marginal damage from a unit of pollution could change over time as the stock changes. If damages increase more than linearly with stock and the stock is increasing over time, then the marginal damage per ton of emissions will increase over time. Second, the marginal damage is the present value of a stream of future impacts caused by this single emission. With long-lasting pollutants such as carbon dioxide, the calculation of marginal damages requires looking far into the future. Chapter 6 of WGIII does not explain these important intertemporal aspects of cumulative pollutants. The comparison of outcomes using current climate versus a $2 \times CO_2$ climate on a 1990 economy does not provide much insight concerning how to properly value a marginal unit of pollution today or in the future.

The discussion of damages from stocks which are four or six times greater than current levels is a distracting and largely irrelevant discussion. The emissions which could generate stocks of this magnitude will not occur until the twenty-second century. What is done over the next thirty years will have virtually no impact on that outcome. Discussing the implications of emissions that will take place in the twenty-second century only obscures the purpose of the valuation, which is to explain what the world will receive from reducing emissions in the near term.

Another source of confusion in the chapter is the discussion of secondary pollutants. The authors argue that greenhouse-gas abatement policies likely will reduce the emissions of other fossil fuel pollutants such as sulfur dioxide, nitrogen dioxide, and particulates. The authors add the benefits of reducing these pollutants to the value of eliminating a ton of carbon. But including secondary effects in the calculation of the marginal damages of emissions of a specific pollutant leads to many problems. First, the value of carbon would vary depending on which secondary pollutants were removed. For example, some abatement policies might remove only carbon whereas others might remove carbon and sulfur dioxide. One would get the odd result of choosing to remove a high-valued ton of carbon not because of the carbon but because of the sulfur dioxide that went with it. Second, one would confuse policies about carbon and policies controlling secondary pollutants. Countries with tight controls on secondary pollutants, such as the United States, could claim they have low-valued carbon emissions whereas other countries with no controls on secondary pollutants would have high-valued carbon emissions. This implies that carbon policies should focus on the most damaging tons of carbon emissions first, that is, the tons from (developing) countries which do not control secondary pollutants. But this conclusion mistakenly confuses carbon and secondary pollutant policies.

Including secondary pollutants in the valuation of a ton of carbon confuses a marginal ton of carbon emissions with an entire carbon-abatement policy. If an abatement policy reduces tons of other pollutants, this benefit of the policy should be taken into account. If a policy introduces a new externality, this added cost should be attributed to the policy. In both of these cases, however, changes in other externalities should have no effect on the value placed on a ton of carbon emissions.

Of course, if one compares alternative energy policies, it is important to consider all externalities. The authors quantify the benefits from reducing fossil fuel consumption associated with the reduction in secondary pollutants from fossil fuels. They must also quantify the increase in externalities associated with any new energy sources which replace fossil fuels. Alternative energy sources such as hydroelectric, nuclear, and renewable energy all involve externalities. Even some programs to reduce energy consumption, for example, by making homes more airtight, have health effects. These externalities must also be measured and included when evaluating energy programs; and they are not likely to be trivial. Fossil fuel pollutants, at least in the OECD, are already heavily controlled, so they do not cause extensive damage any more. The magnitude of damages associated with a heavy reliance on renewable or nuclear energy could well be larger than the elimination of the remaining amounts of secondary pollutants (see Rowe and others 1995).

EMPIRICAL METHODOLOGY

From an economics point of view, one of the most controversial aspects of Chapter 6 of the WGIII report is its empirical estimation of impacts. The authors were not expected to develop original empirical estimates. Unfortunately, the literature available at the time the report was written was insufficient to determine global damage estimates. Further, the existing empirical studies were all based on climate scenarios far more severe than those expected today. Consequently, the authors were forced to sift through this literature and make a judgment concerning their best guess of the global-warming impacts. In this task, the authors have fallen short. The report does not appear to take into account recent changes in scientific forecasts (Houghton and others 1996; Watson and others 1996), and it resists the evolution of methods and findings in the impact literature. Ironically, the report takes the results of older, more primitive studies as well-taken and focuses most of its criticism on the newer state-of-the-art research.

One serious flaw in the empirical impact work is its inconsistency with the most recent science reports of the IPCC (Houghton and others 1996; Watson and others 1996). The typical climate scenario tested in the empirical impact studies cited in the report is a 4.5°C warming, often with reduced summer precipitation, a one-third reduction of runoff, and a sea-level rise of one meter. Even the 1990 IPCC report (Houghton and others 1990) views these climate scenarios as the most severe plausible scenarios. The most recent IPCC report (Houghton and others 1996) estimates that climate will warm by only 2°C by 2100, with precipitation increases and a sea-level rise of 0.5 meters. Although the authors mention that the climate predictions have become more moderate, they curiously state that "the net effect is hard to predict." It is not clear why the authors cannot predict that a large climate change will cause much more damage than a small climate change. In any case, they have not adjusted earlier impact studies to account for these milder scenarios but instead report them at face value.

The authors also assume that climate change will cause expected hurricane damages (see Tables 6.4, 6.5, and 6.8). In contrast, recent science reports (Houghton and others 1996) continue to be unwilling to predict that severe storms are expected to increase as a result of warming. The authors could have pointed out that increases in storms and climate variability would lead to significant damages. However, there is no justification for including storm damage as an expected effect of global warming. The authors' inclusion of impacts from storms and other severe climate scenarios which are now considered unlikely suggests that the IPCC impact numbers are not expected but upper bound estimates.

The authors insist on minimizing the importance of carbon fertilization, claiming it will have a small effect, despite the extensive empirical evidence to the contrary. First, the authors insist that the effective doubling of greenhouse gases entails CO_2 concentrations of 440 ppmv (22% higher than current levels). According to scientific analysis, however, CO_2 concentrations will reach 440 ppmv by 2030 (Houghton and others 1996). By 2100, when temperatures are expected to have increased 2°C, CO_2 concentrations could be as high as 700 ppmv (Houghton and others 1996). The authors have clearly understated the likely magnitude of carbon dioxide in the atmosphere over the next century. They then cite some ecological results which question the magnitude of fertilization effects in natural ecosystems and use them as justification for arguing there will be little carbon fertilization anywhere. That is, they use uncertainties raised about carbon fertilization in natural systems as justification for eliminating known results found for managed systems. By assuming minimal carbon fertilization, the authors appear to be substituting their opinion for a large empirical literature in agronomy and forestry (for a recent review, see Helms and others 1996). Given that the authors have no natural-science expertise, it is not clear what authority they use to dismiss these scientific findings.

Although the authors cite the importance of adaptation, they almost entirely ignore it when evaluating impact studies. They do not criticize a single study which has omitted adaptation. They avoid noting that more recent studies which have included adaptation, such as Kaiser and others (1993), Mendelsohn and others (1994), and Mendelsohn and Neumann (1998), consistently estimate smaller damages. This should imply that previous studies that omitted adaptation overestimate damages. Instead, the authors merely observe "that adaptation is unlikely to completely offset the effects of climate change." By underplaying both carbon fertilization and adaptation, the authors give the impression that world food markets will be at risk from even mild warming. The authors appear to be unaware that most recent agricultural studies report that mild warming will not disrupt world food markets (see Helms and others 1996; Schimmelpfennig and others 1996).

Although Chapter 6 mentions that recent scientific studies suggest forests will do well in a warmer, wetter, CO_2-enriched world (see Watson and others 1996; VEMAP 1995), the authors base all their empirical estimates on scenarios where there is extensive loss of forest cover because of climate change. The authors are aware that these forest losses are based on earlier science and are in direct conflict with newer ecological results, but the forest impacts reported in Tables 6.4, 6.5, and 6.8 are all damages. Once again, the impacts are inconsistent with the most recent science.

The authors also fail to mention the importance of intertemporal adaptation for climate-sensitive sectors with large long-lived capital stocks such as forestry and coastal resources. It is difficult to change large capital stocks quickly. Rapid climate change, which might dictate a restructuring of these stocks, would require adjustments over several decades. Static equilibrium comparisons of outcomes before and after climate change undervalue this important dynamic process and tend to overstate damages. Sohngen and Mendelsohn (1998), for example, find that dynamic adjustments make a huge difference in evaluating the timber market consequences of intertemporal climate-change scenarios. Yohe and others (1996) find that sea-level rise adjustments last for many decades because of the slow turnover rate of the housing stock. By including adaptation through a series of protection and abandonment decisions, Yohe and others calculate that the present value of sea-level rise damages will be an order of magnitude smaller than previously thought.

Chapter 6 does a good job of identifying sectors of the economy and the quality of life which could be damaged by climate warming. No plausible negative impacts are known at this time which were not covered in the chapter. However, the chapter is not as careful in covering beneficial impacts. The only beneficial consequence of warming mentioned in the chapter is the reduction of heating costs and possible agricultural improvements in Russia. The possibility that warming would be good for summer activities such as outdoor recreation is conspicuously overlooked. The fact that people have been moving to warmer climates throughout the United States is overlooked. The possibility that ecosystems would generally flourish in a warmer, wetter, CO_2-enriched world is overlooked. The conspicuous omission of benefits skews the chapter and especially the empirical conclusions.

Although it is important to include nonmarket impacts, there is little empirical foundation upon which to estimate the magnitude of some of these impacts. For example, careful analyses identifying how many and which species would be lost from climate change have not been done. The authors merely postulate that 6,000 plants might be lost or that the lost species from climate change might be worth what people state they would pay for a specific megafauna. These are rather arbitrary assumptions upon which to assign damages of $4 billion to $8 billion per year to a single country. If the damages to endangered species are anywhere near this magnitude of importance, it is disturbing that not a single country has yet bothered to fund research which would quantify the risks to species and the concern of the public.

The authors' conclusion that warming would reduce human amenities has little empirical foundation either. Although people may not like heat waves, hedonic wage studies (see especially Hoch and Drake 1974) consistently indicate that Americans prefer warmer climates. This is also consistent with the observed migration of Americans to the warmer parts of the United States and the lower wages in these regions. The empirical evidence suggests warming will have largely beneficial, not harmful effects on human amenities, at least in the United States.

Another highly controversial estimate in Chapter 6 concerns health effects. The chapter cites estimates such as 23,000 deaths a year in the OECD, 115,000 deaths in the non-OECD, and 215,000 deaths worldwide. Although it is plausible, especially in developing countries, that climate change could affect human health through heat stress, cold stress, starvation, and altering the conditions for vector-borne diseases, there is little evidence that OECD residents would be vulnerable to warming. Heat stress appears to be more sensitive to short-term weather patterns than mean temperatures. The very data sets used to demonstrate heat stress reveal that heat-stress deaths are lower in the southern United States than in the northern United States. Vector-borne diseases are climate-sensitive, but according to the models which predict disease potential, they should already be present in the OECD. The fact that malaria and other vector-borne diseases are not already in the OECD suggests that effective controls to prevent these diseases are already in place. Over time, as they develop, even non-OECD countries may become less sensitive to vector-borne diseases. By the time warming appears, many developing countries may have sufficient incomes to largely control vector-borne diseases.

The magnitude of the chapter's air and water pollution damage estimates is also suspect. Although the authors mention that pollution cannot be valued using abatement costs, the analysis by Titus (1992) of air and water pollution effects makes precisely this assumption. Titus assumes that abatement will have to increase to offset any changes in ambient concentrations caused by climate change. This would require large increases in the abatement of air pollutants which are precursors to ozone because warming would allow greater ozone formation given current levels of hydrocarbons and nitrogen oxides. Similarly, with shrinking runoff in rivers, water pollution would have to be reduced to keep the same concentration of pollution in rivers. The use of abatement costs to measure damages, however, is incorrect and, in the case of these two programs, may greatly overestimate the damages because of the high cost of additional abatement and questions about the magnitude of benefits historically achieved in these programs.

SYNTHESIS

Chapter 6 of the WGII report makes important contributions to the climate-change debate. First, it properly argues that assessing the damages from climate change is a critical component of determining the best social policy toward this problem. Second, the chapter identifies the aspects of the economy and of the quality of life which might be most sensitive to climate change. Third, the chapter notes that the impacts are not likely to be shared equally and alerts countries to investigate what would actually happen within their borders. It is highly likely that the hotter a country is today, the more they will be damaged by further warming and that cooler countries will actually benefit. Fourth, the chapter begins the difficult but essential discussion of environmental values in determining impacts across the globe.

However, this chapter on social costs also has drawbacks. It does not capture the dramatic change in the understanding of climate change that natural and social scientists have made over the last five years. Changes in climate science imply that CO_2 doubling is likely to entail only moderate climate change. These more moderate scenarios will likely lead to dramatically reduced damages. Ecologists have revised their earlier estimates of the impacts of climate change. Ecosystem models now predict that a mildly warmer, wetter, CO_2-enriched climate scenario will be more productive than the current climate. Ecosystem effects are more likely (although not universally) to be beneficial given these new predictions. Economic analyses, such as Mendelsohn and Neumann (1998), reveal that human systems will respond to climate change and adapt to the new conditions. These adaptations will consistently reduce the damages predicted from climate change and enhance the benefits. A complete accounting of climate effects includes many benefits as well as damages. The chapter falls short of including these important new insights.

Because the chapter has failed to take into account the factors above, the estimates in the chapter should be treated as an upper bound of what might happen. The expected value of impacts is not likely to be as harmful as the chapter implies. Net impacts on the world economy are likely to be quite small (near zero) with offsetting effects across sectors and across regions. Colder regions are likely to enjoy benefits; warmer regions are likely to be damaged. Agriculture and forestry are likely to benefit, but energy, coastal structures, and water systems will most likely be hurt. Although even less is known about the impact of climate change on the quality of life, offsetting effects will likely occur here as well. People in temperate and polar climates will likely gain from mild warming while people in the subtropics and tropics will likely be worse off. People who enjoy winter activities like skiing will be hurt whereas people who enjoy summer activities will benefit. Bioproductivity will likely increase but some endangered species will likely be lost. The net effect of these changes is not likely to be as harmful as the chapter implies.

The value of climate impacts, however, remains highly uncertain. The only sectors which have been carefully quantified are market impacts in the United States. The American results can be reasonably extrapolated to other countries with similar climates, income, and development. Much less is known about market impacts in tropical and subtropical countries which begin with different climates, incomes, and economic development. Further, nonmarket impacts are just beginning to be quantified. Health studies continue to be limited to analyzing potential effects rather than predicted impacts. There have been no attempts to date to value ecological changes from climate change. The existing estimates are poor extrapolations from studies done of other issues. The only published study of aesthetic damages is over twenty years old. Because more research has not yet been done, impact estimates are uncertain.

Additional research on impacts is needed to reduce the range of uncertainty. Careful studies which include adaptation have yet to be done for most nonmarket goods, especially species loss, health, and amenities. Additional research is also needed on the economies of developing countries. It may be reasonable to take American studies and extrapolate the results to other OECD countries which share the same income level, technology, and temperate climate. We have much less confidence in extrapolating American results to developing countries with much lower incomes, imperfect markets, primitive technologies, and tropical climates. In order to make credible global impact estimates, empirical studies must be conducted of the impact of climate change on the market economies of developing countries and nonmarket effects throughout the world.

REFERENCES

Adams, R., B. McCarl, K. Segerson, C. Rosenzweig, K. Bryant, B. Dixon, R. Connor, R. Evenson, and D. Ojima. 1998. The Economic Effect of Climate Change on U.S. Agriculture. In *The Economic Impact of Climate Change on the Economy of the United States*, edited by R. Mendelsohn and J. Neumann. Cambridge: Cambridge University Press.

Bruce, J., H. Lee, and E. Haites. 1996. *Climate Change 1995: Economic and Social Dimensions of Climate Change*. Intergovernmental Panel on Climate Change. Cambridge: Cambridge University Press.

Cline, W. 1992. *The Economics of Global Warming*. Washington, D.C.: Institute of International Economics.

Cummings, R., D. Brookshire, and W. Schulze. 1986. *Valuing Environmental Goods: Assessment of the Contingent Valuation Method*. Ottawa, New Jersey: Rowman and Allanheld.

Esty, D. and R. Mendelsohn. 1998. Moving from National to International Environmental Policy. *Policy Analysis* (forthcoming).

Falk, I. and R. Mendelsohn. 1993. The Economics of Controlling Stock Pollution: An Efficient Strategy for Greenhouse Gases. *Journal of Environmental Economics and Management* 38:213–18.

Fankhauser, S. 1995. *Valuing Climate Change: The Economics of the Greenhouse*. London: Earthscan.

Helms, S., R. Mendelsohn, and J. Neumann. 1996. The Impact of Climate Change on Agriculture. *Climatic Change* 33:1–6.

Hoch, J. and J. Drake. 1974. Wages, Climate and the Quality of Life. *Journal of Environmental Economics and Management* 1:268–95.

Houghton, J. T., G. J. Jenkins, and J. J. Ephraums, eds. 1990. *Climate Change: The IPCC Scientific Assessment*. Cambridge: Cambridge University Press.

Houghton, J. T., L. G. Meira Filho, J. Bruce, Hoesung Lee, B. A. Callander, E. Haites, N. Harris, and K. Maskell, eds. 1994. *Climate Change 1994: Radiative Forcing of Climate Change and an Evaluation of the IPCC IS92 Emission Scenarios*. Cambridge: Cambridge University Press.

Houghton, J. T., L. Meira Filho, B. Callander, N. Harris, A. Kattenberg, and K. Maskell, eds. 1996. *Climate Change 1995: The Science of Climate Change*. Intergovernmental Panel on Climate Change. Cambridge: Cambridge University Press.

Kaiser, H., S. Riha, D. Wilkes, and R. Sampath. 1993. Adaptation to Global Climate Change at the Farm Level. In *Agricultural Dimensions of Global Climate Change*, edited by H. Kaiser and T. Drennen. Delray Beach, Florida: St. Lucie Press.

Maddison, D. 1994. *The Shadow Price of Greenhouse Gases and Aerosols*. London: CSERGE, University College.

Mendelsohn, R. and J. Neumann, eds. 1998. *The Economic Impact of Climate Change on the Economy of the United States*. Cambridge: Cambridge University Press.

Mendelsohn, R., W. Nordhaus, and D. Shaw. 1994. The Impact of Global Warming on Agriculture. *American Economic Review* 84(4):753–71.

Nordhaus, W. 1991. To Slow or Not to Slow: The Economics of the Greenhouse Effect. *Economic Journal* 101:920–37

Nordhaus, W. 1994. *Managing the Global Commons: The Economics of Climate Change*. Cambridge, Massachusetts: MIT Press.

Peck, S. and T. Teisberg. 1992. CETA: A Model for Carbon Emissions Trajectory Assessment. *Energy Journal* 13:55–77.

Rosenzweig, C. and M. Parry. 1992. *Climate Change and World Food Supply*. Washington, D.C.: U.S. EPA.

Rowe, R., C. Lang, L. Chestnut, S. Bernow, and D. White. 1995. *New York State Environmental Externality Cost Studies*. Dobbs Ferry, New York: Oceana Publications.

Samuelson, P. 1964. The Pure Theory of Public Expenditure. In *Collected Scientific Papers of Paul Samuelson* (1966), edited by J. Stiglitz. Cambridge, Massachusetts: MIT Press.

Scheraga, J., N. Leary, R. Goettle, D. Jorgenson, and P. Wilcoxen. 1993. Macroeconomic Modelling and the Assessment of Climate Change Impacts. In *Costs,*

Impacts, and Benefits of CO$_2$ Mitigation, edited by E. Kaya, N. Nakicenovic, W. Nordhaus, and F. Toth. IIASA Paper Series, CP93-2. Laxenburg, Austria.

Schimmelpfennig, D., J. Lewandrowski, J. Reilly, M. Tsigas, and I. Parry. 1996. *Agricultural Adaptation to Climatic Change: Issues of Long-run Sustainability.* Report AER-740. Washington, D.C.: Economics Research Service, USDA.

Smith, J. and D. Tirpak. 1989. *The Potential Effects of Global Climate Change on the United States: Report to Congress.* EPA–230–05–89–050. Washington, D.C.: U.S. EPA.

Sohngen, B. and R. Mendelsohn. 1998. The Economic Effect of Climate Change on U.S. Timber Markets. In *The Economic Impact of Climate Change on the Economy of the United States,* edited by R. Mendelsohn and J. Neumann. Cambridge: Cambridge University Press.

Titus, J. 1992. The Costs of Climate Change to the United States. In *Global Climate Change: Implications, Challenges, and Mitigation Measures,* edited by S. Majumdar, L. Kalkstein, B. Yarnal, E. Miller, and L. Rosenfeld. Philadelphia: Pennsylvania Academy of Sciences.

Tol, R. 1995. The Damage Costs of Climate Change Toward More Comprehensive Calculations. *Environmental Resource Economics* 5:353–74.

VEMAP Members. 1995. Vegetation/Ecosystem Modeling and Analysis Project: Comparing Biogeographic and Biogeochemistry Models in a Continental-Scale Study of Terrestrial Ecosystem Response to Climate Change and CO$_2$ Doubling. *Global Biogeochemical Cycles* 9:407–37.

Watson, R., M. Zinyowera, R. Moss, and D. Dokken. 1996. *Climate Change 1995: Intergovernmental Panel on Climate Change Impacts, Adaptations, and Mitigation of Climate Change.* Cambridge: Cambridge University Press.

Yohe, G., J. Neumann, P. Marshall, and H. Ameden. 1996. The Economic Cost of Sea Level Rise on U.S. Coastal Properties. *Climatic Change* 32:387–410.

Comment

Climate-Change Damages

Richard S. J. Tol

The discussion around the "social cost chapter," Chapter 6 of the Second Assessment Report of Working Group III of the Intergovernmental Panel on Climate Change (Pearce and others 1996) has long been unscholarly and intemperate. I am very pleased that the discussion is moving away from that. I thank Professor Mendelsohn for his critical review of the social-cost chapter and for providing an alternative set of estimates of the impacts of climate change.

Mendelsohn correctly points out that greenhouse gases are stock pollutants. Chapter 6 indeed focuses on the impact of $2 \times CO_2$ on the current economy, but it does also discuss, in Section 6.6, how to derive the marginal value of emissions, including, *inter alia*, the aspects raised by Mendelsohn in his review. Section 6.6 is admittedly short and may easily be overlooked. The reason for its brevity is twofold. First, there is little discussion on the subject in the literature. Second, since calculating marginal damages requires a considerable amount of modeling, a more extensive discussion would have been more appropriately placed in the chapter on integrated assessment (Chapter 10). However, communication between lead author teams was not always easy.

Defective communication could also be observed between the social-cost chapter and the chapter on equity (Chapter 3: Banuri and others 1996). The forced separation between valuation and justice is one of the causes of the controversy that arose about Chapter 6 (see, for example, Bruce 1995, 1996; Courtney 1996; Fankhauser and Tol 1995; Grubb 1996;

DR. RICHARD S. J. TOL is a researcher at the Institute for Environmental Studies, Vrije Universiteit, Amsterdam.

Masood 1995; Masood and Ochert 1995; Meyer 1995; Meyer and Cooper 1995; Pearce 1995; Sundaraman 1995). Mendelsohn points out another unfortunate way in which material was divided over the different chapters. The social-cost chapter discusses the secondary benefits of fossil fuel abatement because the literature on estimates of secondary benefits is methodologically more akin to the literature on the social costs of climate change than to the literature on the costs of emissions reduction. Nonetheless, the section on estimates of secondary benefits would have been more appropriately placed in Chapter 9, on emissions-reduction costs (Hourcade and others 1996).

Mendelsohn states that pollution cannot be valued using abatement costs. It is true that the expenditures on greenhouse-gas-emissions abatement are not necessarily a proper indication for the costs of climate change. However, the additional costs of meeting air- and water-pollution standards may be used as a rough indication of the impact of climate change on these sectors. If pollution standards were set on the basis of a cost-benefit analysis, standards would shift with climate change, and Mendelsohn would be right to say that the cost of meeting the current standard is an inappropriate measure of the impacts of climate change. Pollution standards are seldom based on cost-benefit analysis, however. If we assume lexicographic preferences, that is, that all is well if pollution falls below the standard and dramatically bad if it rises above, then the standard remains the same and the additional costs of meeting it are a proper measure. Reality lies somewhere in between. It is convenient to assume constant pollution standards.

The newer set of market-impact estimates put forward by Mendelsohn (Mendelsohn and others 1996a, b; Mendelsohn and Neumann 1998) is not as novel in methodology or outcome as he would like us to believe. For example, the supposed innovation of Yohe and others (1996) is perfect foresight—rational adaptation to sea-level rise. The idea can be traced back to Yohe (1991), while Fankhauser (1994) was the first to implement it on a broad scale. [Fankhauser was initially unaware of Yohe's work, as was Yohe of Fankhauser's in his applied work for the United States (Fankhauser and Yohe, personal communication).] The study by Yohe and others (1996) is more detailed and the empirical basis is stronger. The Fankhauser study includes all OECD countries (instead of just the United States) and takes wetlands into consideration. Although optimal adaptation is an improvement over earlier assumptions, one may wonder whether perfect foresight and rational behavior are realistic assumptions when it comes to coastal defense.

The methodological novelties that are present are not necessarily a step forward. For instance, the study by Mendelsohn and others (1995) on agriculture (underlying Mendelsohn and others 1996a, b) assumes future adaptation to climate to be equal to observed adaptation. The Rosenzweig and Parry (1994) study on agriculture (underlying Tol 1995) assumes arbitrary adaptation of various degrees, which is inferior but does yield insight into the transient period between equilibria and adaptation policies. The Ricardian approach of Mendelsohn and others (1995) allows them to consider only U.S. agriculture in isolation from other sectors and the rest of the world. Rosenzweig and Parry (1994) consider changes in agricultural prices and international agricultural trade, albeit in a rudimentary form.

Table 1 compares the aggregate market impacts for $2 \times CO_2$ as estimated by Fankhauser (1995), Tol (1995), and Mendelsohn and others (1996b). The first two studies are the basis for the regional estimates reported in Chapter 6; they are here corrected for PPP-exchange rates (Fankhauser and Tol 1997). The results are not widely disparate, given that these are best guesses with wide (though unassessed) uncertainty bounds. Note that market impacts are only a subset of total impacts; the best-guess nonmarket impact for the whole is in the order of magnitude of a loss of 1% to –2% of GDP equivalent. Note that the figures in Table 1 are not readily comparable because the assumed climate change, the set of impact categories, and the definition of the regions differ among the studies. No systematic biases result from this for any of the studies. The Fankhauser and Tol estimates are based on purchasing power parity exchange rates, whereas the estimates of Mendelsohn and others are based on market exchange rates. Mendelsohn and others thus underestimate non-OECD, "southern," and global damages. These would increase if the damages were "equity weighted," for instance, because of risk aversion or inequality aversion (compare to Fankhauser and others 1997).

In his review, Mendelsohn often wonders why a particular study was not taken up in the social-cost chapter or in the aggregate-damage esti-

Table 1. Market Impacts (percent of GDP), 2.5° C Warming.[a]

Region	Fankhauser	Mendelsohn	Tol
OECD	0.77	–0.17	0.27
Non-OECD	0.67	0.03	0.76
North[b]	0.60	–0.23	–0.06
South[b]	1.02	0.17	1.66
World	0.72	–0.08	0.52

a. Mendelsohn assumes a 2.5° C rise in global mean temperature to take place in 2060, whereas Fankhauser and Tol assume this to happen in 2050. Note that only Tol has damage depending on the rate of climate change. In all three cases, vulnerability is assumed as in 1990.

b. The countries subsumed under the labels "North" and "South" differ among the assessments. Fankhauser's north is OECD plus the former Soviet Union. Mendelsohn's north is North and Central America, Europe, and Oceania for damages, and the United States, other OECD, and the former Soviet Union for GDP. Tol's north is OECD plus Central and Eastern Europe and the former Soviet Union.

Source: Fankhauser and Tol 1997, after Fankhauser 1995, Tol 1995, and Mendelsohn and others 1996b.

mates. To start with the first question, a literature survey can only include what is known by its authors or pointed out by its referees. More importantly, because of the political trouble the social-cost chapter found itself in, its authors opted for a strict interpretation of the terms of reference of the IPCC. As a consequence, the survey only extends to March 1995 and to a large extent excludes the (then) gray literature.

The question as to why more recent material did not find its way into the aggregate-damage estimates hints at an important shortcoming of the literature on the economic evaluation of climate-change impacts. This shortcoming is present, though not readily visible, in the pre–March 1995 literature. It has grown more visible since, partly as a result of the IPCC process; and solutions are now being discussed.

The shortcoming is this: the earlier studies focus on impact estimates (that is, numbers), not on impact estimation (that is, methods). Earlier studies tend to (1) discuss the underlying literature on case studies of climate-change impacts for various countries, regions, and sectors; (2) hint at rules for valuation, reconciliation, extrapolation, and aggregation; and (3) present comprehensive damage estimates. The published literature only hints at how the end result was obtained. The methods are not necessarily written down in internal reports or stored in the analysts' computers, let alone subject to open discussion. This is particularly the case for the reconciliation of various, often conflicting, pieces of information for one particular country or impact category, and for scaling up from case studies to comprehensive estimates.

As a result, existing aggregate-impact assessments are not readily amended by new evidence. A new study does not become an additional observation in a well-organized database. A new finding does not become a changed parameter in a widely accepted model. Instead, new results require tedious reinterpretation by the original analyst and a new round of peer review. In my opinion, this is the prime reason why the aggregate damages in the IPCC chapter are slightly outdated. It is also something that figures high on my research agenda, and I think it should be high on others' research agendas as well, if we want the Third Assessment Report to be more internally consistent.

REFERENCES

Banuri, T., K.-G. Mäler, M. Grubb, H. K. Jacobson, and F. Yamin. 1996. Equity and Social Considerations. Chapter 3 in IPCC 1996.

Bruce, J. P. 1995. Impact of Climate Change. *Nature* 377:472.

———. 1996. Purpose and Function of IPCC. *Nature* 379:108–9.

Courtney, R. S. 1996. Purpose and Function of IPCC. *Nature* 379:109.

Fankhauser, S. 1994. Protection vs. Retreat: The Economic Costs of Sea Level Rise. *Environment and Planning* A 27:29–319.

———. 1995. *Valuing Climate Change: The Economics of the Greenhouse.* London: EarthScan.

Fankhauser, S. and R. S. J. Tol. 1995. A Recalculation of the Social Costs of Climate Change: A Comment. Occasional Paper Series. *The Ecologist.*

———. 1997. *The Social Costs of Climate Change: The IPCC Second Assessment Report and Beyond. Mitigation and Adaptation Strategies for Global Change.* Cambridge: Cambridge University Press.

Fankhauser, S., R. S. J. Tol, and D. W. Pearce. 1997. The Aggregation of Climate Change Damages: A Welfare-Theoretic Approach. *Environmental and Resource Economics* 10:249–66.

Grubb, M. 1996. Purpose and Function of IPCC. *Nature* 379:108.

Hourcade, J. C., K. Halsneas, M. Jaccard, W. D. Montgomery, R. Richels, J. Robinson, P. R. Shukla, and P. Sturm. 1996. A Review of Mitigation Cost Studies. Chapter 9 in IPCC 1996.

IPCC (Intergovernmental Panel on Climate Change). 1996. *Climate Change 1995: Economic and Social Dimensions of Climate Change. The Contribution of Working Group III to the Second Assessment Report of the Intergovernmental Panel on Climate Change.* Edited by J. P. Bruce, H. Lee, and E. F. Haites. Cambridge: Cambridge University Press.

Masood, E. 1995. Developing Countries Dispute Use of Figures on Climate Change Impact. *Nature* 376:374.

Masood, E. and A. Ochert. 1995. U.N. Climate Change Report Turns Up the Heat. *Nature* 378:119.

Mendelsohn, R., R. Adams, J. M. Callaway, B. Hurd, B. McCarl, W. Morrison, K. Segerson, J. Smith, B. Sohngen, and G. Yohe. 1996a. *The Market Impacts of Climate Change in the U.S.* (draft).

Mendelsohn, R., W. Morrison, M. E. Schlesinger, and N. G. Andronova. 1996b. A Global Impact Model for Climate Change (draft).

Mendelsohn, R. and J. O. Neumann, eds. 1998. *The Impacts of Climate Change on the U.S. Economy.* Cambridge: Cambridge University Press.

Mendelsohn, R., W. D. Nordhaus, and D. Shaw. 1995. The Impact of Climate on Agriculture: A Ricardian Analysis. *American Economic Review* 84(4): 753–71.

Meyer, A. 1995. Economics of Climate Change. *Nature* 378:433.

Meyer, A. and T. Cooper. 1995. A Recalculation of the Social Costs of Climate Change. Occasional Paper Series. *The Ecologist.*

Pearce, D. W. 1995. Valuing Climate Change. *Chemistry and Industry* 1024.

Pearce, D. W., W. R. Cline, A. N. Achanta, S. Fankhauser, R. K. Pachauri, R. S. J. Tol, and P. Vellinga. 1996. The Social Costs of Climate Change: Greenhouse Damage and the Benefits of Control. Chapter 6 in IPCC 1996.

Rosenzweig, C. and M. L. Parry. 1994. Potential Impact of Climate Change on World Food Supply. *Nature* 367:133–38.

Sundaraman, N. 1995. Impact of Climate Change. *Nature* 377:472.

Tol, R. S. J. 1995. The Damage Costs of Climate Change Toward More Comprehensive Calculations. *Environmental and Resource Economics* 5:353–74.

Weyant, J., O. Davidson, H. Dowlatabadi, J. Edmonds, M. Grubb, E. A. Parson, R. Richels, J. Rotmans, P. R. Shukla, R. S. J. Tol, W. R. Cline, and S. Fankhauser. 1996. Integrated Assessment of Climate Change: An Overview and Comparison of Approaches and Results. Chapter 10 in IPCC 1996.

Yohe, G. W. 1991. Uncertainty, Climate Change and the Economic Valuation of Information: An Economic Methodology for Evaluating the Timing and Relative Efficacy of Alternative Responses to Climate Change with Application to Protecting Developed Property from Greenhouse Induced Sea Level Rise. *Policy Sciences* 24:245–69.

Yohe, G. W., J. Neumann, P. Marshall, and H. Ameden. 1996. The Economic Costs of Sea Level Rise on U.S. Coastal Properties. *Climatic Change* 32:387–410.

Comments

Climate-Change Damages

John Reilly

Overall, Robert Mendelsohn identifies most of the main issues associated with the IPCC Working Group III chapter Social Costs of Climate Change (Pearce and others 1996; hereafter referred to as Chapter 6). His basic message is that Chapter 6 is biased toward portraying damages as more severe than a careful reading of the current technical literature would indicate. Recent studies of potential damages support his contention that earlier literature likely underestimated adaptation potential in market sectors. However, the further implication of his message—that we can be fairly certain damages are less than the 1.5% to 2.0% of GNP reported in Chapter 6 as the best-guess central estimates of available studies—likely overstates the certainty we should have with regard to damage estimates. This paper will, first, briefly identify Mendelsohn's principal claims for why he believes the chapter overestimates damages, evaluating the evidence to support these claims. It will then address a set of issues which together support the contention of this paper, that a 90% or 95% confidence limit on future climate damages would easily contain both the estimates of Chapter 6 and Mendelsohn's argument for much lower damage estimates or possible benefits.

EVIDENCE FOR HIGH-DAMAGE BIAS

In Mendelsohn's view, the reader of Chapter 6 is left with a biased view of climate damages for numerous reasons. The chapter (1) fails to cite

JOHN REILLY is a senior scientist in the Joint Center for Climate Change Science and Policy at the Massachusetts Institute of Technology.

more recent literature which shows lower damages; (2) sharply criticizes more recent literature that, in general, has a richer technical foundation while citing with little criticism older literature or damage estimates based on far weaker foundations; (3) fails to fully consider adaptation potential; (4) omits beneficial impacts of climate change; (5) cites damage estimates based only on doubling and quadrupling of greenhouse gases; and (6) bases damage estimates on climate scenarios that show much higher climate sensitivity than is currently believed likely or than is consistent with current science. He also criticizes the chapter for failing to take explicit account of the stock nature of the greenhouse-gas problem. He commends the IPCC and the chapter authors for taking on the controversial issue of summarizing impacts in economic-value terms. He also defends the economic principle by which the chapter reaches different values for the potential loss of human life in different regions of the world—which was one of the most controversial aspects of the 1996 IPCC reports. In general, Mendelsohn's conclusion that the chapter leaves the reader with an impression that damages may be higher than the most recent literature suggests is somewhat persuasive, although he overstates the case.

With regard to the currency of the literature, the chapter is fairly effective in citing recent literature; 100 of about 250 citations have a publication date of 1993 or later. Given that writing for the chapter began in 1993, this feature of the reference list is fairly strong evidence of an ongoing effort to keep the chapter current with new literature even through its various stages of review. Mendelsohn's concerns derive from a lack of consistency between sections of the chapter that review damage studies for sectors/regions and those that attempt to comprehensively add up damages with the intent of generating a global damage function over time that could be used in an appropriately formulated economic cost-benefit analysis. The final summary of the chapter relies more on the latter studies.

Among the aggregate studies, there is little evidence of a trend toward lower damages as Mendelsohn argues. The chapter's Table 6.4, which reviews such estimates for the United States (where there is a longer time series of estimates), actually shows the lowest estimate to be the first one published (Nordhaus 1991) and the highest estimate to be the second one published (Titus 1992). Estimates since then have been between these two, alternately lower or higher than the preceding.

To the extent that the chapter relies on aggregate studies for its main conclusions, Mendelsohn's contention that the estimates lack currency has some validity even when the cited study has a very current date. For example, the Fankhauser (1995) compilation uses estimates for agricultural damages from a study published in 1992 which was in fact based on work conducted in 1989 and 1990 to support the 1990 IPCC assessment.

Thus, the data cited above on publication dates may overstate the currency of the science. Given the state of impact assessment, it was likely unavoidable that this chapter would rely on aggregate studies that synthesize literature.

The evidence Mendelsohn cites for recent studies that were not included are studies that have looked only at the United States or specific sites within the United States. The difficulty for Chapter 6 was how to update global estimates on the basis of such studies. As Reilly and others (1994) show, the economic effect on a country, predicted on the basis of assessing that single country as if climate change were not global, can be of the wrong sign compared with an economic assessment that assumes agricultural production potential is simultaneously changing everywhere. This is simply the result of changing prices and a country's position as an exporter or importer. Thus, the work of Mendelsohn, Nordhaus, and Shaw (1994) and similar work should be considered as an estimate of the initial productivity shock to the agricultural sector in the United States rather than a measure of the economic effect, unless one argues that world prices for commodities will not change. As an approach to measure adaptation potential with the ceteris paribus assumption that all prices are held constant, the approach has merit. Chapter 6's authors could have suggested that all damages should be reduced by some factor. The 1995 paper by Mendelsohn and others uses a uniform 2.8°C temperature and 8% increase in precipitation for the entire United States. This makes it hard to know whether the overall net impacts measure they generate is different from other results because of adaptation or because the climate scenario they use does not show the temperate midcontinental drying that exists in some global circulation model (GCM) scenarios. For the MINK study (Easterling and others 1993), Chapter 6 states that adaptation cuts losses by about one-third. This compares with adaptation that cuts losses by about 10% (against impacts without adaptation or CO_2 fertilization) in the Reilly and others (1994) global study. Thus, this is some evidence that, at least in the United States, a broader consideration of adaptation could reduce damage estimates.

The approach by Mendelsohn and others (1995) further assumes that relative prices among crops do not change. They claim, among other reasons, that their approach generates more positive effects because it includes the fact that climatic conditions for many small-acreage but high-valued crops are improving in many areas while conditions for major grains, which account for much of crop acreage, are worsening. Thus, they are capturing this further adjustment, which studies assessing a few major crops have not captured. Again, if markets reestablish a new equilibrium, expansion of high-valued crops would be limited by a fall in their price as supply increased. Assuming that prices do not change would tend to overestimate adaptation. Thus, in the literature available at the time Chapter 6 was written, actually reviewed and documented evidence for much more adaptation potential was limited.

More research has appeared since publication of the IPCC report that supports Mendelsohn's contention of greater adaptation potential. To construct a meaningful comparison of studies, Schimmelpfennig and others (1996) worked with authors of a number of the agricultural studies cited by Mendelsohn to rerun models using comparable base conditions and comparable climate scenarios. The climate scenarios were GISS, GFDL, and UKMO, which have global average temperature increases of 4.2°C, 4.0°C, and 5.2°C, respectively. A comparison of yield estimates based at two U.S. sites shows that, even for these extreme scenarios, adaptation can more than offset losses in some cases (Table 1). Estimates by Darwin and others (1995) similarly show much greater adaptation potential globally (Table 2). Neither of these tables includes the CO_2 fertilization effect which, at an effective doubling of CO_2, based on experiments under controlled conditions, could increase average crop yields by 10% to 15%.[1]

Table 1. Yield Impacts for Various Climate Scenarios.

	Without adaptation	With adaptation
Maize	−31	+3
Soybeans	−28	+15
Winter wheat	−20	−4

Note: These yield impacts are calculated with farm-level adaptation consistent with economic optimization and without adaptation, no CO_2 fertilization effect (percentage change, average of three climate scenarios and two sites).

Source: Summarized from Table 2.1, Schimmelpfennig and others 1996.

Table 2. Percentage Changes in Global Supply and Production of Cereals by Climate-Change Scenario.

	Supply		Production	
World	No adaptation	Land use fixed	Land use fixed	No restrictions
GISS	−22.6	−2.4	0.2	0.9
GFDL	−23.5	−4.4	−0.6	0.3
UKMO	−29.3	−6.4	−0.2	1.2
OSU	−18.6	−3.9	−0.5	0.2

Note: Changes in supply represent the additional quantities firms would be willing to sell at 1990 prices under the alternative climate. Changes in production represent changes in equilibrium quantities, under new equilibrium prices. The results are based on 2×CO_2 equilibrium scenarios for four climate models, those developed at the Oregon State University (OSU), Geophysical Fluids Dynamics Laboratory (GFDL), Goddard Institute of Space Studies (GISS), and United Kingdom Meteorological Office (UKMO).

Source: Darwin and others, 1995.

With regard to the overly sharp criticism of recent literature and failure to criticize literature based on weaker foundations, I do not read the chapter as introducing a particular bias as Mendelsohn suggests. The richer technical foundation for some areas such as agriculture simply provides more opportunity to discuss some of the strategies for estimating impacts, illustrating some of the strengths and weaknesses of different approaches. Unfortunately, for most sectors other than agriculture and sea level the research is far less mature. Thus, the implications of international trade for damage estimates have not been explored for most sectors as they have for agriculture. Nor have researchers pursued two fundamentally different strategies in such a way that one could compare and contrast results to better understand potential biases in each. In agriculture, the empirical approach of considering analogous regions (ultimately developed as a cross-section econometric problem) has been developed alongside integrated biophysical-economic models parameterized on the basis of scientific understanding of plant growth and explicit characterization of the decision process a farmer undertakes throughout a growing season with specified technical-economic adaptation options. These multiple approaches provide the opportunity to confirm results or suggest avenues for further research that could resolve differences.

In contrast to Mendelsohn's reading, I find the chapter to be fairly frank about the lack of empirical research or methodological development for impact areas like amenities and ecosystems. With regard to amenity impacts, after reviewing three studies, the authors conclude that it is not possible to say whether climate change will increase or decrease amenity value. The impacts could be large, and the direction of impact would tend to vary across regions; thus losses in some areas would be balanced by gains in others. The first paragraph of the discussion on ecosystem losses ends with the statement, "there is a serious need for conceptual and quantitative work in this area"; and toward the end of the section the authors conclude that "monetary estimates of ecosystem damages through climate change are invariably *ad hoc*." In discussing how estimates were generated, the chapter is fairly explicit. For example, in reviewing Cline's estimates of amenity value, it states, "assuming people are willing to pay 0.25% of their income to avoid this (increased heat spells)." In the ecosystem and biodiversity-loss section, the chapter takes the reader through the logic of where the authors got estimates for the average value of a lost species and how this was multiplied by an estimate of number of species lost to arrive at a total value. The frank discussion of the fact that values were simply assumed in some cases combined with the warning of their *ad hoc* nature does more to undermine economics as a science than to mislead policymakers about the level of damages. The lack of methodology leaves little more to say than that the values were assumed or developed in an ad hoc manner. If there is an error, it is in reporting empirical estimates at all given their quality. Given that the empirical estimates are included in the aggregate estimates and that valuation of these aspects of climate change is an important component of integrated assessment—discussed in Chapter 10 of the IPCC report (Weyant and others 1996)—the frank discussion is useful.

Mendelsohn's concern about using out-of-date climate scenarios is a possible source of serious bias. Many detailed studies of climate change have used the GISS, GFDL, and UKMO climate scenarios; some have used the OSU scenarios which, as illustrated in Figure 1, are quite high relative to the Working Group I expectations for climate change through 2100. The IPCC transient climate paths were generated from a highly simplified climate model that is tuned to judgments about climate sensitivity and the role of the oceans. No full-scale GCMs have actually been run in a transient mode with emissions paths as projected by economic models. Thus, there is not a completely defensible way to estimate transient impacts consistent with the IPCC climate scenarios when the impact models require spatially disaggregated climate projections. Working Group II of the IPCC actually attempted to work with the climate-science community to generate some reference scenarios for 2025 and 2050, but Working Group I climate scientists would not agree to stand behind the scenarios.

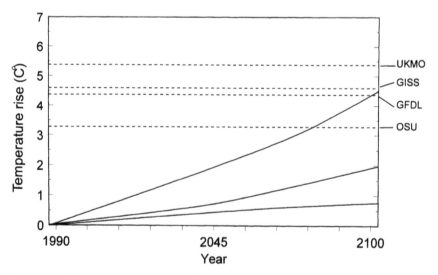

Figure 1. Global Mean Temperature Rise.

Note: Projection of global mean temperature from 1990 to 2100 for three climate sensitivities (4.5, 2.5, and 1.5 °C) and a median emissions scenario including uncertainty in future aerosol concentrations. Increasing aerosols (e.g., sulfur, also from coal burning) are estimated to have a cooling effect (after IPCC 1996). See Table 2, p. 246, for identification of acronyms.

Chapter 6 provides Table 6.5, with estimates of damages under a 2.5°C warming. While many other tables cover damages under different conditions, this table enables the reader to relate damages to roughly the mean estimate of climate change expected to occur by 2100. Thus, Mendelsohn's claim that the chapter reports estimates inconsistent with current understanding of the science is not completely accurate. In order to get estimates for a 2.5°C warming, however, the authors had to scale down damages because most impact studies have been conducted for larger temperature increases. This major problem for damage estimates has no obvious solution short of improving the research on which these estimates are based. The difficulty increases as one tries to move toward explicitly recognizing the stock nature of the problem of generating the current value of control—which Mendelsohn argues is essential. To do so, one must describe the entire path of damages. Reilly and Richards (1993), fitting linear and quadratic damage functions to damage "data points" for a CO_2 doubling and quadrupling, show that these different approaches to the damage function could significantly affect estimates of the current level of control, largely because of what these different representations meant for damages over the next fifty years. Yet, all the "data points" are for levels of climate change that are not expected to occur for nearly one hundred years. Given the lack of research on nearer-term damages, the more scientifically justified approach may be to simply present the estimates for $2 \times CO_2$ scenarios and locate them in time/probability space to be as consistent with Working Group I as possible. Any scaling to more or less climate change is purely ad hoc given the lack of empirical evidence. Mendelsohn reports a view that a temperature change of up to 2.5°C benefits global agriculture, further illustrating that arbitrary scaling cannot even be bounded to exclude the possibility of net benefits over a significant period of warming. However, the evidence for agricultural benefits up to a 2.5°C warming had not yet been published, which limited the ability of Chapter 6's authors to include these results.

Mendelsohn's criticism of the chapter's inconsistency with Working Group I is not easily remedied. Actually having evidence on transient impacts is inconsistent with Working Group I because of its failure to generate spatially disaggregated transient climate scenarios. Given this lack, if the standard for economic studies of the current value of climate-change control is consistency with current science, then none of the economic studies that purport to estimate the current value of control or to compare the costs and benefits of control can pass the test.

The Reilly and Richards (1994) result and Mendelsohn's claims for agricultural benefits for some amount of climate change both provide strong support for Mendelsohn's observation that Chapter 6 misallocated space by addressing the very long-term (one hundred to two hundred year out) effects. Efforts to estimate impacts over the next ten to one hundred years have a far higher priority for informing decisions that must be made over the next decade. There probably is room for some thought about the chance for runaway climate change, thresholds, and truly catastrophic events that could occur over the longer term. However, any thinking about climate damages one hundred and two hundred years in the future requires more fundamental thinking about the nature of the relationship between humanity and the planet than can be captured in crop models and valuation methods that intrinsically reflect the value of marginal changes.

Summarizing the evidence for a bias toward overestimating damages, some of the choices the chapter made may contribute to an impression of greater damages than the current literature indicates. Some of the more important literature to support this current view was, however, not published or widely available until after the IPCC report was published.

ARE WE TOO CERTAIN ABOUT CLIMATE-CHANGE DAMAGES?

A more serious bias generated by reading Chapter 6 or the Mendelsohn critique is the impression they leave with the reader that economists have come close to developing a reliable estimate of the damages of climate change. Neither the state of climate projections nor the state of economic assessment of impacts should leave us with much confidence in any of the estimates of damages that have been developed. The research in this area has made progress, but attaching much weight to specific estimates for policy purposes is probably premature. The research does support the conclusion that, simply because a warming of 3°C or 4°C over one hundred years may be unprecedented for the earth as a natural system, this change does not necessarily equate to unprecedented impacts on the economy and society. As currently understood, this amount of climate change would have far from catastrophic effects.

A number of issues lead to significant uncertainty. The following points focus on why, even with the evidence Mendelsohn presents, damages as high or higher than those presented in Chapter 6 cannot easily be ruled out.

It is true, as Mendelsohn points out, that climate scientists are not prepared to say that hurricanes will increase in number or intensity (although there is some consensus and evidence that rainfall will be more intense). But this is simply a statement of lack of knowledge and understanding of why hurricanes form rather than a statement that intensity and frequency will not increase. Realistically, we must consider that hurricane intensity and frequency may increase or decrease. Given this significant uncertainty, the biggest concern with estimates of increased storm damage in Chapter 6 is that they portray more certainty than is justified.

Are there catastrophes—Where are the tails of the distribution? Chapter 6 mainly reports efforts to describe central tendencies of damage estimates. The policymakers summary (IPCC 1996) argues that "consideration of risk aversion, and application of the precautionary principle provide rationales for action beyond no regrets." For risk aversion to be a justifiable basis for additional abatement at the global level the risks would have to be substantially greater than the many risks world economic activity normally faces. If they are not substantially greater, climate change should be treated as just another element in the portfolio of activities society undertakes. The possibility of catastrophic consequences occasionally enters climate-change discussions. The melting of the West Antarctic Ice Sheet, a runaway greenhouse effect from the release of methane hydrates in permafrost or shallow coastal regions, and changes in the oceans' conveyor belt are events that have been suggested and in some cases examined and dismissed as highly improbable. Without more explicit consideration of how damaging these events would be and what the chance is of their occurrence (however small), it is unclear whether a significant risk-aversion premium should be associated with climate-change policy. Open discussion of low-probability, high-consequence events remains an issue because of the difficulty of communicating without causing undue alarm. Given our limited understanding of earth systems, the bigger problem is trying to deal with true surprise rather than more conventional risk problems where the distribution of outcomes can be reasonably described.

Caution about translating a reduction in the global temperature estimate as necessarily a reduction in damages is warranted because other factors and features of climate change that are moderating the global temperature increase may actually lead to larger damages. For example, the major reason the 1995 IPCC science report estimated a reduced global temperature increase for 2100 (compared with the previous IPCC report) was the consideration of the cooling effect of aerosols. This offsetting cooling effect occurs largely in northern temperate regions (for example, United States, Europe, and China) because that is where current and future aerosol emissions are likely to occur. But these are also the areas where, for example, agriculture is frequently estimated to benefit from warming. So the effect of aerosols may be to reduce the benefits of warming that would accrue to northern hemisphere temperate regions while not reducing the damages that would occur in tropical and subtropical regions. Thus, global damages would not only not decrease because the global temperature estimate decreased but would actually be higher. Decreased insolation due to aerosol haze could contribute further to decreased crop growth.

Mendelsohn looks on ecosystem models showing greater biomass productivity as indicative of climate-change benefits, but more biomass growth in ecosystems may not necessarily be a good thing. More biomass in a forest translates into more wood products and more carbon sequestration. Apart from the fact that ecosystem models remain primitive in their representation of dynamic adjustment to transient climate forcing, for the range of nonmarket benefits of ecosystems (species preservation, recreation) it is not clear that benefits necessarily increase with biomass production. The viewing quality of ecosystems may decline if there is significant forest dieback or weedy species invade existing ecosystems; rare species may disappear; and human disease vectors could expand in ecosystems, creating health risks for recreators.

The adjustment process and the potential that adjustment could increase costs have not been factored into many of the recent analyses Mendelsohn cites. Thus, while earlier studies are perhaps overly pessimistic, these recent studies may be overly optimistic. If we could trust that the rate of global climate change indicates the rate of local climate change, then we might be able to comfortably dismiss adjustment costs for market sectors. The local smoothness of climate change can hardly be assumed, particularly for precipitation—which can change dramatically for local areas if the storm track changes by fifty or one hundred miles. No realistic transient climate scenarios have been developed, making it impossible to rule out a pattern of climate change where local areas exhibit stability for some period of time and then change rapidly over a few years. Such a pattern could impose serious adjustment costs even with accurate forecasting and forward-looking behavior.

There remains a real question of the ability of agents to detect and successfully adapt to climate change given the huge variability in weather from year to year. For purposes of adaptation, research must consider how much of the adaptation we estimate is keyed to direct observation of climate/weather (I look out the window, it is raining, I take my umbrella) versus that which requires an expectation about weather (it was an unusually hot and dry summer, should I install air conditioning/irrigation equipment). Mendelsohn argues that static comparisons overestimate damages because agents are forward looking and will begin planning ahead by changing tree-planting practices and building coastal protection now in anticipation of climate change. Except for sea-level rise, where no one disputes the direction of change, the ability to look forward may be severely limited. For tree planting, water resource management, and ecosystem protection, climate science cannot indicate confidently if a region will get wetter or drier, whether storms will become more or less severe, or whether temperature increases will occur mainly in the winter or summer or night or day. Reilly and Thomas (1993) lay out some aspects of this problem, considering the issue of autonomous adjustment versus nonautonomous adjustment, and recognizing the difference between adjustments that are fully issues of short-run adjustment (where asset fixity is not an issue) versus those where the long-run behavior must be considered (where asset fixity is an issue).

The concept of asset fixity is readily transferable, conceptually, to problems of ecosystem adjustment and to consumer preferences. Should preferences be represented as a stock variable: exposure to specific conditions (I like cool Wisconsin summers because I grew up there, and if they are suddenly taken away, I would be worse off; but with exposure to warm Washington, D.C., summers I might learn to like it hot)? Or are climatic/ecosystem/recreation preferences independent of experience and dependent only on largely exogenous factors such as income? In either case, the research evidence is limited.

Misguided responses to climate change are possible. Just as a world agreement on climate change will not likely be implemented in an optimal manner (not all countries will participate, inefficient regulatory/command-and-control policies will be used instead of incentive-based policies), private or governmental agents will not likely respond efficiently to the forces of climate change. Countries that lose comparative advantage in agricultural exports may erect trade barriers to protect market share. Existing conflicts on water rights within and among countries may prevent the efficient allocation of water if it becomes scarcer. Coastal areas may fail to take precautionary measures as sea level rises. Irrigation investment may expand in areas that should be abandoned. Properties may continue to be rebuilt in areas that have become increasingly flood prone.

IN SUMMARY

In general, Mendelsohn raises important issues and sets out the case for why climate-change damages may be lower than represented in Chapter 6. New evidence available after the IPCC report was published supports his contention. A major difference among the camps of those who think the impacts of climate change will be small and those who see them as large is a set of implicit assumptions that underlies the methodologies chosen to investigate this problem. These are assumptions about the ease of adaptation, ability to detect climate change, ability of the economic and political system to react correctly to the signals presented, and beliefs about where responsibility rests for those who fail to adapt. Until more hard empirical evidence is presented on these issues, disagreement about potential damages will likely continue.

In setting out arguments to support the case for higher damages, the intention of this paper was to communicate the fact that, even in those areas of climate-change damage that have been comparatively well researched, many unresolved issues remain. If pushed to give a best guess of the net global damages at 2.5°C, I would likely place the estimate below the 1.5% to 2.0% of world GDP cited in Chapter 6 as "*best-guess central estimates.*" My subjective 90% confidence interval for net damages would include a significant likelihood of net benefits but would not exclude damages as large or larger than those cited in the chapter.

ENDNOTES

1. For C3 crops, the average experimental result is that a doubling of CO_2 from 300 to 600 ppm would increase yields by 30%. Only one-half to two-thirds of the trace gas doubling may be due to CO_2, thus the yield for C3 crops at effective doubling may only be 15% to 20%. C4 crops may only increase yields by 7% at doubling (3% to 5% at equivalent doubling). C3 crops are about two-thirds and C4 crops one-third of world crop production.

2. Effects differ by plant group. The two important plant groups for agriculture are C3 and C4, which refers to the carbon pathway in the photosynthetic mechanism. The C3 photosynthetic mechanism is so called because the first compound into which CO_2 is incorporated is a compound with three carbon atoms. Most crop species are C3, including wheat, rice, barley, oats, potatoes, and most vegetable crops. C4 plants have a unique CO_2-concentrating mechanism within their leaves, accomplished by CO_2 first being incorporated into a four-carbon compound. These plants tend to be of tropical origin and include corn, sorghum, sugar cane, and many tropical grasses.

REFERENCES

Darwin, R., M. Tsigas, J. Lewandrowski, and A. Raneses. 1995. *World Agriculture and Climate Change: Economic Adaptation.* Report No. AER-709. Washington, D.C.: Economic Research Service, USDA.

Easterling, W. E. III, P. R. Crosson, N. J. Rosenberg, M. McKenney, L. A. Katz, and K. Lemon. 1993. Agricultural Impacts of and Responses to Climate Change in the Missouri-Iowa-Nebraska-Kansas (MINK) Region. *Climatic Change* 24:23–61.

Fankhauser, S. 1995. *Valuing Climate Change. The Economics of the Greenhouse.* London: Earthscan.

IPCC (Intergovernmental Panel on Climate Change). 1996. *Climate Change 1995: Economic and Social Dimensions of Climate Change. The Contribution of Working Group III to the Second Assessment Report of the Intergovernmental Panel on Cli-*

mate Change. Edited by J. P. Bruce, H. Lee, and E. F. Haites. Cambridge: Cambridge University Press.

Mendelsohn, R., W. Nordhaus, and D. Shaw. 1994. The Impact of Climate on Agriculture: A Ricardian Approach. *American Economic Review* 84(Sept.):753–71.

Nordhaus, W. 1991. To Slow or Not to Slow: The Economics of the Greenhouse Effect. *Economic Journal* 101(407):920–37.

Pearce, D. W., W. R. Cline, A. N. Achanta, S. Fankhauser, R. K. Pachauri, R. S. J. Tol, and P. Vellinga. 1996. Social Costs of Climate Change: Greenhouse Damages and the Benefits of Control. Chapter 6 in IPCC 1996.

Reilly, J., N. Hohmann, and S. Kane. 1994. Climate Change and Agricultural Trade: Who Benefits, Who Loses? *Global Environmental Change* 4(March):24–36.

Reilly, J. and K. Richards. 1993. An Economic Interpretation of the Trace Gas Index Issue. *Environmental and Resource Economics* 3:41–61.

Reilly, J. and Chris Thomas. 1993. *Toward Economic Evaluation of Climate Change Impacts: A Review and Evaluation of Studies of the Impact of Climate Change.* MIT-CEEPR 93–009WP, Center for Energy and Environmental Policy Research, MIT, June.

Rosenzweig, C. and M. L. Parry. 1994. Potential Impacts of Climate Change on World Food Supply. *Nature* 367(13 Jan.):133–38.

Schimmelpfennig, D., J. Lewandrowski, J. Reilly, M. Tsigas, and I. Parry. 1996. *Agricultural Adaptation to Climate Change: Issues of Longrun Sustainability.* Report No. AER-740. Washington, D.C: Economic Research Service, USDA.

Titus, J. G. 1992. The Costs of Climate Change in the United States. In *Global Climate Change: Implications, Challenges, and Mitigation Measures,* edited by S. K. Majumbar, L. S. Kalkstein, B. Yarnal, E. W. Miller, and L. M. Rosenfeld. Easton: Pennsylvania Academy of Sciences.

Weyant, J. and others. 1996. Integrated Assessment of Climate Change: An Overview and Comparison of Approaches and Results. In IPCC 1996.

Comment

Climate-Change Damages

William R. Cline

The central thrust of Mendelsohn's review of the IPCC Working Group III Chapter 6 on social costs of climate change is that the authors overstate these costs, in part because their references are out of date. As one of the authors of the chapter in question, I continue to believe that it is a closer representation of reality than the minimal-damage diagnosis offered by Mendelsohn.

His first critique is that the chapter fails to take account of more benign climate projections in the more recent scientific work. While the recent IPCC scientific estimates anticipate somewhat slower realized warming because of aerosol masking, they do not change the earlier IPCC warming parameter (1.5°C to 4.5°C for a doubling of atmospheric carbon equivalent); and progress in reducing urban pollution in the coming century could reduce aerosol masking (Wigley and Raper 1992). Moreover, as the chapter notes, the recent IPCC scientific results show relatively faster warming in the northern hemisphere, leaving the baseline there little changed from the earlier transient projections, while at the same time increasing the emphasis on regional variation and surprises. Indeed, my reading of the evolving scientific work is that it has increasingly stressed how suddenly extreme shifts in climate can occur (for example, as recorded in ice cores; see, for example, Anklin and others 1993).This suggests that the recent science might more appropriately be seen as the basis for greater rather than lesser concern. Finally, many of the estimates in the underlying literature surveyed in Chapter 6 (including Cline 1992)

WILLIAM R. CLINE is Deputy Managing Director and Chief Economist at the Institute of International Finance, Washington D.C.

already took into account the fact that the 1990 IPCC central parameter for global warming (2.5°C for CO_2 doubling) was below the median for the leading general circulation models, and had already adjusted some of the EPA and other estimates accordingly.

Similarly, Mendelsohn implies that Chapter 6 fails to recognize that more recent economic studies show less severe damage from global warming. As examples, he cites recent studies on sea-level rise showing sharply lower damages through a strategy of planned obsolescence. However, it is erroneous to extrapolate to total sea-level damage costs the recalculated damages associated with coastline structures, because by far the bulk of sea-rise damages stems from recurrent annual losses of dryland and wetlands rather than the once-and-for-all capital losses associated with structures.

On agriculture, Chapter 6 does take into account Mendelsohn's own recent Ricardian model (Mendelsohn, Nordhaus, and Shaw 1994), which purports to find small damages or even benefits to agriculture from global warming, but regards this study as unpersuasive (for reasons set forth in detail in Cline 1996).[1] Mendelsohn seeks to cast doubt on the literature on agriculture by treating it as having "robot" assumptions of nonadaptation, yet Chapter 6 takes into account the MINK and Rosenzweig-Parry studies which explicitly incorporate adaptation. Similarly, the chapter recognizes the differentiation between more severe agricultural impacts in developing countries and only modest, or initially even positive, effects in industrial countries as identified in the recent work by the international team led by Rosenzweig and Parry. But it also notes that the same study finds severe impacts if carbon fertilization is less than the study's central assumption (550 ppm), or if precipitation increases at a lower rate. On the latter point, Mendelsohn takes issue with my use of 440 ppm as the appropriate steady-state CO_2 concentration for doubling of preindustrial atmospheric equivalent. Yet this can be calculated directly from the IPCC scientific analysis of the contribution of noncarbon gases to radiative forcing (as set forth in Cline 1992, 92–93).[2]

Mendelsohn makes much of Chapter 6's supposed failure to recognize that global-warming effects have dynamic properties derived from the cumulative stock nature of atmospheric buildup and criticizes the chapter's use of comparative static conditions as between present and doubled-concentration climates. However, the authors do in fact take pains to describe the cumulative, dynamic nature of the process; and several of the shadow prices reported in the chapter for different future periods are derived, as noted explicitly in the chapter (Pearce and others 1996, 214–15), from dynamic optimization models—including my own simulations of Nordhaus's DICE model in which I apply an alternative discounting approach.

Mendelsohn dismisses the chapter's reference to much larger damages over the time scale of two to three centuries. This is curious, considering that the principal economic optimization models have increasingly explicitly incorporated time scales of two centuries or more rather than cutting off the analysis in 2100 or before. It is also curious given Mendelsohn's own emphasis on the cumulative nature of the problem.

In forestry, he suggests that recent studies imply larger growth rather than damages. However, focusing on the ultimate steady state for forests misses the point of major losses during an initial period of perhaps two centuries when the pace of forest migration on the poleward perimeter cannot keep up with the pace of dieoff on the equatorward perimeter. Mendelsohn, in fact, commends a warmer, wetter world as conducive to agriculture and vegetation but seems not to recognize the problem that increased precipitation will be concentrated in the highest latitudes, leaving continental midlatitudes and today's grainbelts drier rather than wetter (Rind and others 1990).

Mendelsohn criticizes the inclusion of storm damage. The WGIII Chapter 6 authors did have to make a decision on whether to report these estimates from the literature despite continued uncertainty in the IPCC scientific review. We were impressed by the view of the lead author of the IPCC scientific work, Houghton (1994), that hurricanes would in fact increase with warming. Also, given the considerably higher dollar damages associated with more recent hurricanes compared with the literature, we judged that the existing estimates might not be misleading as expected values (considering that there is little if any scientific sentiment that hurricane incidence and intensity will *decline*). In this regard, we already have a market test: within private industry, it is the insurers that have been the most concerned about global warming because of the risks of large losses.

On sea-level rise, the estimates surveyed in Chapter 6 frequently use one meter as a benchmark. The authors were aware that this exceeds the most recent central estimate of fifty centimeters by 2100. However, we argued that use of the one-meter measure is probably less misleading because it compensates somewhat for the failure to incorporate the fact that sea level will keep rising for centuries even if atmospheric concentration stops rising in 2100.

On secondary benefits of reduced pollution resulting from antigreenhouse measures, Mendelsohn's critique fails to recognize that the central estimates of damage as a percent of GDP reported in the chapter do not include these effects. His critique also ignores the chapter's explicit point that "each pollutant should be taxed in proportion to the environmental damage it causes" (Pearce and others 1996, 218), just as he suggests.

It is gratifying that Mendelsohn commends the chapter authors for dealing directly with the difficult issue of valuation of impacts on human life. His attack on the estimates for species loss is easy enough to make, given the conceptual difficulties in this area. However, is zero (the implicit value if the category is simply omitted) a better estimate? The chapter authors were actually somewhat surprised that under completely different methods the orders of magnitude of these losses came out relatively similar.

Mendelsohn's central conclusion is that "a 2.5°C warming is predicted to have only a small impact on market economies." For the various reasons set forth here—especially potential future aerosol unmasking, the concentration of additional precipitation where it does little good, and lesser carbon fertilization when noncarbon greenhouse gases are taken into account—I disagree with this diagnosis even for the central estimate. What I find surprising, however, is that Mendelsohn simply ignores what is the principal concern of most scientists and even most economists looking at this issue: the noncentral cases—upper-end warming parameters and especially damages from catastrophic events not included in the central estimates. In any economic analysis that finds abatement attractive on the basis of a cost-benefit analysis of the central case (as, with modest allowance for risk aversion, is found in Cline 1992), there is no burden of identifying the value of catastrophic risks; they simply strengthen the main case. For anyone who wishes to argue that damages are too small to warrant much action, however, as Mendelsohn apparently does, the burden of attempting to take catastrophic risk into account surely cannot be set aside.

ENDNOTES

1. Nor am I persuaded by the reply of Mendelsohn and Nordhaus (1996) on the same issue. They continue to assume away the problem of scarce availability of irrigation in a greenhouse future. If their cropland-weighted results are used rather than their crop-revenue-weighted results (because of the key role of irrigation for the latter), their most recent calculations actually increase expected U.S. agricultural damage from climate change (doubling of carbon equivalent) to $703 billion capital loss, or $35 billion annual loss. This is about twice the damage estimate in Cline (1992) and six times the original estimate in Mendelsohn, Nordhaus, and Shaw (1994). The higher damage estimate results from applying actual general circulation model estimates for U.S. climate conditions rather than the global averages.

2. Even in terms of the central Rosenzweig-Parry estimates, severe cutbacks in production are found for developing countries where there is less flexibility for

adaptation. Although Mendelsohn concludes that "most recent agricultural studies report that mild warming will not disrupt world food markets," few would consider benchmark 2.5°C warming "mild," and lesser warming seems broadly irrelevant to the fundamental debate.

REFERENCES

Anklin, M. and others. 1993. Climate Instability during the Last Interflacial Period Recorded in the GRIP Ice Core. *Nature* 364:202–7.

Cline, William R. 1992. *The Economics of Global Warming*. Washington, D.C.: Institute for International Economics.

———. 1996. The Impact of Global Warming on Agriculture: Comment. *American Economic Review* 86(5):1309–11. (A comment on Mendelsohn, Nordhaus, and Shaw 1994.)

Houghton, J. 1994. *Global Warming: The Complete Briefing*. London: Lion Book.

Mendelsohn, Robert, and William Nordhaus. 1996. The Impact of Global Warming on Agriculture: Reply. *American Economic Review* 86(5):1312–15.

Mendelsohn, Robert, William D. Nordhaus, and Daigee Shaw. 1994. The Impact of Global Warming on Agriculture: A Ricardian Approach. *American Economic Review* 84(4):753–71. (See also Cline 1996.)

Pearce, D. W., W. R. Cline, A. N. Anchata, S. Fankhauser, R. K. Pachauri, R. S. J. Tol, and P. Vellinga. 1996. The Social Costs of Climate Change: Greenhouse Damage and the Benefits of Control. Chapter 6 in *Climate Change 1995: Economic and Social Dimensions of Climate Change*, edited by James P. Bruce, Hoesung Lee, and Erik F. Haites. Cambridge: Cambridge University Press.

Rind, D., D. Goldberg, J. Hansen, C. Rosenzweig, and R. Ruedy. 1990. Potential Evapotranspiration and the Likelihood of Future Drought. *Journal of Geophysical Research* 95(D7):9,983–10,004.

Wigley, T. M. L. and S. C. B. Raper. 1992. Implications for Climate and Sea Level of Revised IPCC Emissions Scenarios. *Nature* 357:293–300.

9

Integrated Assessment Modeling of Climate Change

Charles D. Kolstad

Although the importance of social and economic factors to climate-change policy has long been recognized, the focus of the written output of the Intergovernmental Panel on Climate Change (IPCC) historically has been the physics and biology of the problem. This slant changed in 1996 with the release of the report of Working Group III of the IPCC on the socioeconomic dimensions of the problem (IPCC 1996). An important tool for understanding the pros and cons of various climate policies is the integrated assessment model, a computer representation of the economics, physics, and other aspects of the problem that are deemed important to formulating policy. Chapter 10 of the Working Group III review concerns integrated assessment models of climate change, a number of which have emerged over the past decade.[1] This paper reviews that chapter and discusses just what an integrated assessment model is and for what purpose it is used.

The IPCC's integrated assessment chapter was assembled by a very distinguished group of analysts. As might be expected, the chapter organizes the integrated assessment field well and provides insight and a sense of structure for what has become a crowded field of models. Perhaps the most constructive approach to commenting on such a review is to take a fresh look at the subject that has become known as integrated assessment modeling and to offer a few observations that can perhaps focus thinking about future research agendas. Before doing that, I will briefly review the integrated assessment chapter.

CHARLES D. KOLSTAD is a professor in the Department of Economics and the Bren School of Environmental Science and Management at the University of California–Santa Barbara.

THE IPCC CHAPTER ON INTEGRATED ASSESSMENT

The authors of the integrated assessment chapter (Chapter 10) have approached the difficult task of summarizing integrated assessment by focusing on the principal results that have emerged from integrated assessment. In addition, they briefly discuss the history of integrated assessment and close by providing an evaluation of strengths, limitations, and future directions for integrated assessment.

Integrated Assessment Models

The integrated assessment chapter defines integrated assessment as analysis which involves multiple disciplines and seeks to inform policy and decisionmaking rather than to advance knowledge for its own sake. Consequently, disciplinary research that informs policy does not qualify as integrated assessment; nor does multidisciplinary work that serves to advance knowledge rather than inform policy. The multidisciplinary nature of integrated assessment involves representing the physical, ecological, economic, and social aspects of climate change. Integrated assessment serves the policy process in three ways: by helping assess specific policies to control climate change, by providing a systematic framework for structuring present knowledge about climate change, and by giving perspective to the costs of climate change by comparing them to the costs of other human needs.

For the most part, integrated assessment involves computational models. However, the authors of the integrated assessment chapter explicitly embrace a broader definition of integrated assessment than just computer models. They include in their definition panels of experts, informally linked disciplinary models, interdisciplinary research teams, and even individuals with competence in multiple disciplines who may write essays integrating multiple dimensions of a problem. The authors point out that although there are advantages to using numerical models for integrated assessments, the disadvantages (relative to teams of experts, for example) are often overlooked. These disadvantages include the inability to realistically represent social, political, or other aspects of a problem because of the necessity of representing these aspects in a particular mathematical formulation with computational constraints.

For historical perspective, the authors note that integrated assessment predates the current application to climate change. They report that in the 1970s integrated assessments were conducted for supersonic passenger flight (via teams of experts) and for global development (the "Limits to Growth" debate). The authors also note the integrated assessment work in the 1980s on acid rain. Finally, they point out that climate-change

integrated assessment first emerged in the late 1970s from earlier economic and technical models of energy policy.

Seeking to add to some structure to understanding the diverse set of
integrated assessment models in existence, the authors of Chapter 10
have offered several categories of integrated assessment models. One criterion for categorization is the extent to which a model is "full scale." By
full scale, the authors mean that the model deals with the multitude of
issues that concern the IPCC—the model is all-inclusive. This means that
the model should include detail on diverse aspects of the climate problem, ranging from the multiplicity of greenhouse gases (GHGs), to the
variety of human activities contributing to emissions and impacts, to different ways in which the climate and oceans respond to increased levels
of greenhouse gases. The authors make clear that they view full-scale
models as encompassing as much as possible, including nonglobal environmental problems such as local air pollution.

Another classification of integrated assessment models distinguishes
between policy evaluation models and policy optimization models. Policy
evaluation models are designed to project the consequences of a particular climate policy whereas a policy optimization model is designed to
choose a "best" policy from a set of alternatives, based on a specified policy goal or objective. A policy optimization model may find the best way
of meeting a particular goal (such as stabilizing equivalent CO_2 concentrations at 550 ppm) or may seek to balance the costs and benefits of controlling climate change. In economics, this focus on finding the least-cost
way of achieving an arbitrary environmental target is termed regulatory
cost effectiveness. In contrast, regulatory efficiency would involve finding
the appropriate target as well, a target which would balance the costs of
control with the benefits of control.

Although it may sound as if policy optimization models are simply
policy evaluation models with an additional component that chooses
among alternative policies, in reality policy optimization models tend to
be highly aggregated. In contrast, the policy evaluation models often
include much more detail on the physical effects of climate change on different countries or regions, including effects in nonmarket sectors. However, policy evaluation models often stop short of placing economic values on these impacts, a feature common to policy optimization models.

An additional category of integrated assessment models involves the
treatment of uncertainty. The authors identify two types of policy evaluation models: deterministic and stochastic. They also identify two types of
policy optimization models: deterministic and those which examine
sequential decisionmaking, based on the arrival of information over time.

Of the twenty-two integrated assessment models identified, nearly
two-thirds are policy optimization models. In their review of the charac-

teristics of these models, the authors note that the policy evaluation models tend to have more detail on physical characteristics of climate change, particularly in the realm of impacts and forcings. On the other hand, some policy evaluation models appear to be weak in "socioeconomic dynamics." Two models are identified as involving no economics (at least as characterized by the authors of the integrated assessment chapter); both are policy evaluation models.

Results of Integrated Assessment

Just as the proof of the pudding is in the eating, the worth of integrated assessment can be gauged by the usefulness of the results that have emerged from integrated assessment. Thus one of the central portions of Chapter 10 is the discussion of policy conclusions that have been drawn from integrated assessment.

Accordingly, the chapter lists the following six policy-related results as having emerged from policy evaluation models:

- Identification of land-use summary measures—Using the IMAGE model, "results show that regional demands for land can serve as a surrogate for measuring local land cover changes."
- Uncertainty in global warming potentials—Using the IMAGE model, results show that global warming potentials as calculated by the IPCC may be off by as much as 5% to 10%.
- Importance of sulphate aerosols—Because sulfur emissions are tied to combustion of coal, efforts to reduce carbon emissions may in fact result in warming because of associated reductions in sulfur dioxide emissions.
- Generation of IPCC scenarios—Four IPCC scenarios for emissions (including "Business-as-Usual") were generated using two integrated assessment models.
- Significance of delaying control policies—IMAGE was used to conclude that a ten-year delay in control would result in a minor increase in global mean temperature.
- Evaluation of rate of change of temperature—The Advisory Group on Greenhouse Gases recommended a maximum rate of temperature rise of 0.1°C per decade. An integrated assessment model concluded that all but one of the IPCC emissions scenarios exceed these rates of change.

Different results were identified for the policy optimization models. Although the authors are less explicit about precisely what are the most significant results, four results appear in the narrative of the chapter:

- Optimal current emissions control is modest—Several of the models agree that the costs and benefits of climate change justify current-period (1990 or 1995) reduction in greenhouse-gas emissions in the range of 5% to 10% (relative to no control) with associated carbon taxes on the order of $5 to $10 per ton of carbon.
- The discount rate is important—A number of models report that lowering the discount rate has the effect of substantially increasing the current-period optimal level of control of greenhouse gases.
- Aggressive control is more costly than moderate control for same concentration target—Two models have been used to examine a variety of paths of control for greenhouse-gas emissions in order to meet the same concentration target (500 ppmv in 2100). The models conclude that costs vary greatly. In general, modest reductions in early years followed by more aggressive reductions are cheaper than sharper reductions in the near term.
- Adjustment costs justify current-period emissions control—Several analyses have shown that one reason for adopting more aggressive emissions control in the near term is that society needs time to adjust its level of control. Costs of emissions control may be influenced by experience; and it takes time to turn over the carbon-emitting capital stock. If ignoring adjustment costs calls for significant controls in the future, then costs may be lowered overall by phasing in controls more gradually.

The chapter cites a number of other results, but most of these are results from single models, not the result of a consensus of several models. For instance, one result from Nordhaus is cited: "roughly speaking, the optimal carbon tax doubles when uncertainty is taken into account, and the optimal control rate increases by slightly less than half."

Conclusions and Recommendations

The authors identify five major challenges facing integrated assessment:

- Develop better summary measures of climate change—Such measures are a necessary component for developing better estimates of the damages of climate change.
- Develop better methods for analyzing catastrophes—Methods are poorly developed for dealing with low-probability, high-consequence events, particularly when little is known about the nature and scope of such events.
- Improve representation of developing countries—Most integrated assessments focus on OECD countries.

- Improve model integration and management—As integrated assessment models become larger and larger, the problems of managing these models will increase.
- Increase relevance of model results—Improve the relevance of results of integrated assessment models as well as the way in which results are presented to policymakers.

Chapter 10 closes with a call for diversity. Model diversity is like biodiversity—a good thing. The authors of the chapter acknowledge that different integrated assessment models serve different purposes and that this diversity is a plus. For instance, simple models suggest fruitful directions for future research, the results of which can be used to enhance more complex models. The authors expect that as time passes, the different approaches to integrated assessment will tend to converge.

WHAT IS INTEGRATED ASSESSMENT?

The term integrated assessment (IA) or integrated assessment modeling/models (IAM) is a relatively recent creation, though much of the "discipline" is not new, dating back at least twenty-five years.

The Purpose of Integrated Assessment

Chapter 10 identifies three purposes of integrated assessment: (1) assess climate-change control policies, (2) constructively force multiple dimensions of the climate-change problem into the same framework, and (3) quantify the relative importance of climate change in the context of other environmental and other nonenvironmental problems facing mankind.

Despite this statement, the purpose of IA remains unclear. Do we really believe that policymakers directly use the results of large models, most of which agree on little, as the basis for policy? Do we really believe that IA is shaping the debate on the relative importance of climate change versus, for instance, alleviating poverty in the third world (an activity that may compete for the same pot of money)? Do we really believe that by forcing output from global circulation models (GCMs) into IAMs that we are shaping the course of research into GCMs? With a few exceptions, the results of IAMs rarely find their way out of the modeling community into the policy community.[2] Furthermore, policymakers are rarely willing to make a policy decision on the basis of output from a "black box."

So why do integrated assessment? In my view, integrated assessment has a less lofty though still very significant role to play in the climate-change debate.[3] The primary role of integrated assessment is to analyze complex questions that can only be answered by integrating multiple disciplinary aspects of the climate problem. Such questions might include the following:

- What level of emissions control is consistent with identified damages from climate change and costs of controlling emissions?
- How much precautionary behavior is warranted by the uncertainties associated with climate change and the potential large-scale repercussions from climate change?
- How much can we rely on technological change to extricate us from the climate-change problem?
- What types of research have the highest potential payoff?
- To what extent should emissions control be deferred until we know more about climate change?
- At what time should we move from R&D to aggressive control of the problem?
- What is the loss from delaying action by a few decades?
- What are the policy choices for achieving particular environmental goals, and what are their relative merits?

Typically, such questions are not answered unequivocally, nor is the answer even the most important product of the integrated assessment. For instance, in William Nordhaus's seminal work with the DICE model,[4] one of the most important outputs was the conclusion that only very modest control was justified, contrary to many people's intuition that climate change is an extremely serious problem that needs immediate strong action. Few people have dwelled on the precise amount of control that was emerging from DICE; rather, the result focused the attention of the research community onto looking for a justification for serious control of greenhouse gases. The model result was less interesting for its policy relevance than for its implications for directing future research.

In a paper on the policy relevance of climate research, Rubin and others (1992) also see integrated assessment as providing insights and direction for climate research and policy. They argue that the primary purposes of an integrated assessment are to survey the state of current knowledge and to reach scientifically informed judgments about what is and what isn't known. They emphasize the role integrated assessment plays as a bridge between the scientific community and the policy community.

In several places, the chapter seeks to distinguish integrated assessment from disciplinary research. This distinction is useful but misleading. Disciplinary research is driven by problems that are disciplinary in nature, problems whose answers can be useful in a variety of contexts, including integrated assessment. But integrated assessment is also research, not merely the generation of analyses of hypothetical policies. The process of conducting integrated assessment is a process of identifying basic questions (such as those listed above) and trying to understand the fundamental forces shaping the answers to these questions. More often than not, the process of answering these questions generates a new set of questions. This is very much a process of research with the IAM serving as a vehicle to illuminate the answers to specific questions.

Defining Integrated Assessment Models

The integrated assessment chapter invokes a very broad definition of integrated assessment, encompassing not only computer models but also loose collections of models as well as expert panels. While I have no quarrel with this sweeping definition, for the most part integrated assessment has come to mean numerical (computerized) integrated assessment models. I will focus on this narrower definition of integrated assessment.

The chapter also suggests that any model that incorporates more than one "component" of climate-economy qualifies as an integrated assessment model.[5] In my view, an integrated assessment model is a model that includes both human activity and some key aspects of the physical relationships driving climate change. The emphasis is on the human activities. A general circulation model (GCM) that uses as input a trajectory of anthropogenic GHG emissions would not qualify as an IAM, because it has no human-dimensions component. Similarly, a GCM linked with an ecosystems model would not be an IAM (although I recognize some would disagree with me here). A model of an economy which can estimate the effect of a carbon tax might be useful but would not qualify as an IAM.

Several categories of IAMs are considered in the integrated assessment chapter. One is based on whether or not a model is "end-to-end," meaning it includes the process of generating emissions as well as the damage from emissions, typically framed within the economics paradigm. For most models, this means that emissions cause damage which causes emissions policy which in turn determines emissions—the loop is closed. Another category compares policy optimization and policy evaluation models. The former type of model endogenously computes the optimal level of control of greenhouse gases. The latter evaluates the effect on the economy and/or the environment of a particular policy to control climate change.

An obvious distinction that is not made in the integrated assessment chapter is between IAMs that explicitly model the response of the economy to climate change versus those that are more like accounting models, with rules of thumb regarding how sectors of the economy respond to climate change. While this is not a sharp distinction, the policy optimization models often explicitly represent the process whereby an economy equilibrates supply and demand. Economics commonly uses surplus maximization to simulate how an economy operates. In a competitive model, supply and demand balance where surplus is maximized. With externalities (for example, pollution), surplus maximization simulates how the economy would operate with the externality optimally corrected. Thus optimization is a way of simulating the economy's dynamics and response to particular policies. The fact that optimization is used does not necessarily mean a model is normative.

The basic reason policy evaluation models do not use optimization is that the models are generally too large and complex for optimization to be practical. This also means that it is computationally difficult to balance the forces of supply and demand in the simulation. Thus policy evaluation models of climate change often omit economic decisions or adopt heuristic rules-of-thumb for a number of economic decisions. This is illustrated in the chapter's Table 10.3 summarizing the characteristics of several dozen IAMs, including the extent to which economic decisionmaking is incorporated into the model. Two of the IAMs in the table do not even consider economics in representing socioeconomic dynamics.

One implication of relying on optimization to simulate the operation of an economy is that model size, detail, and structure are limited by the numerical algorithms used to solve optimization problems. A challenge for the future is to bring these two approaches to modeling closer together. Policy optimization models could benefit from including some of the detail found in the policy evaluation models; policy evaluation models could benefit from the economic behavior and decisionmaking implicit in many policy optimization models.

History of IAMs

As suggested in Chapter 10, integrated assessment has a fairly long history in economics and environmental policymaking, though in other arenas than climate change. It would appear that the name "integrated assessment" is relatively new, but the methodology dates back at least to the OPEC oil price rise and embargo of 1973, if not earlier.[6] Much of the work of which I am aware is from the United States, although undoubtedly parallel activity took place in other countries. One of the largest integrated assessment models was the Project Independence Evaluation System (PIES). This was a very large integrated model used for evaluating various energy policies from the viewpoints of economic efficiency, regional impacts, and environmental consequences (among other things). PIES has changed over the years, but the U.S. government has always used an integrating model for energy policy formation. The current model is called the National Energy Modeling System (NEMS).

Also, in the early 1970s, the National Coal Model (NCM) was developed in the United States for the specific purpose of examining air pollution regulations and other policies which might affect the U.S. coal industry. The NCM was widely used throughout the latter part of the 1970s and well into the 1980s to evaluate a variety of environmental policies. In fact, it played an important role in structuring the air quality regulations implemented from the 1977 Clean Air Act Amendments. At issue were the impacts on high-sulfur-coal producers and mining employment from various regulations controlling sulfur emissions.[7] The NCM was subsequently used for the decade-long debate over acid-rain legislation in the United States, culminating in the 1990 act setting up the sulfur-allowance trading system. A key aspect of air pollution regulation vis-à-vis coal in the United States has always been distributional: which regions lose, which win. Policy formulation in this area has relied heavily on the forecasts from the NCM on regional effects of regulations.

During the 1980s, millions of dollars were expended on an integrated assessment of acid rain in the United States—the National Acid Precipitation Assessment Program (NAPAP). After ten years of study, NAPAP issued its integrated assessment report to serve as guidance to policymakers for acid rain control. Unfortunately, as detailed in Rubin (1991), NAPAP was woefully inadequate in influencing policy. The basic problem was that the natural science done for the assessment was not very useful for examining policy. In Europe, the RAINS model was developed to help formulate regulatory policies for controlling sulfur emissions (Alcamo and others 1990).

A wide variety of other integrated assessment models have been developed for various environmental problems. Atkinson and Lewis (1974) created one of the first models to look at different ways of meeting air pollution targets for an urban area (St. Louis). Later, Atkinson and Tietenberg (1982) examined different policies for achieving ambient air quality targets in the same region, showing the efficiency gains associated with different regulatory alternatives. Kolstad (1986) provided a similar analysis on a regional scale, measuring the efficiency of different marketable-permit schemes to control regional air quality. Literally dozens of other such integrated assessment models have been developed over the last two decades.[8]

The point of this short list of nonclimate IAMs is to suggest that what has been learned in the nonclimate arena should not be forgotten in assessing climate-change IAMs. In fact, the work over the past two decades of the Energy Modeling Forum, led by Chapter 10's lead author, John Weyant, documents the extent of activity in policy modeling, much of which can be characterized as integrated assessment modeling. The integrated assessment chapter could benefit from placing integrated assessment modeling of climate change in this broader context.

In my view, two basic lessons can be taken from this experience. First, complex models are of limited usefulness, particularly if their structure is nontransparent.[9] There are two audiences for any model: the academic/ modeling community and the policy community. For either community to use a model, they must understand how the model works and be able to understand its output intuitively. While complex models are potentially understandable, few people have the time or patience to study them intimately. Even a model builder is not always sure why a model is giving a particular result—whether it is an error or some interesting, hitherto poorly understood aspect of the problem. It is important to realize that large is not necessarily the same as complex. The National Coal Model, mentioned above, is structurally very simple but contains a great deal of data—it is large and simple. Thus we can relatively easily "trust" the output of the model. This, along with its regional detail, may explain its usefulness to the policy process.

The second lesson from two decades of integrated assessment is that forecasts per se are less interesting than insights into the problem. The most successful way to use a model is to have the model generate an apparently counterintuitive result which, ex post, can be explained relatively simply, without resorting to model results. This is useful both to the research community and to the policy community. This is one reason it is often difficult to point to a model result that has specifically influenced policy. What happens is that the model assists the analyst to see more clearly the structure of the problem.

The Role of Projection in Integrated Assessment Models

How important is projecting/predicting the future accurately for integrated assessment? Some would say that is the purpose of an integrated assessment model: to pull together the diverse strands of understanding about a problem to generate the best prediction possible about what the future may bring or about the future effects of a specific policy. The problem is that the quest for accuracy often breeds complexity. And complexity breeds opaqueness—which for an integrated assessment model can mean uselessness.

At the other end of the spectrum are the small models of climate change that seek to capture the essence of the climate-change process without adding complexity. Nordhaus's DICE model fits into this category. With such a simple structure, it is possible to communicate the model's operation to others and for others to learn from its use. This is one reason why many analysts have used the DICE model as the starting point for their own examination of specific issues. Table 10.1 in the integrated assessment chapter lists three offspring of DICE, and there are others that are not listed. The problem with simple, elegant models is that they may be too simplified to be useful to policymaking. For instance, DICE aggregates the world into one region and simulates the operation of the world economy with three equations. This is the essence of economic modeling—strip away unnecessary detail and focus on the core structure of the problem. We learn from this exercise but cannot place any "faith" in the numerical results that emerge from the model, at least as a basis for policy. Manne and Richels are midway between simplicity and complexity with their Global 2100 model and its progeny. They have a structurally simple economic sector and a significantly detailed energy sector. Despite this, many would consider their model structure overly simple to form the basis for policy.

Thus this tension remains between detail/realism and simplicity/ transparency. It is not easy to find the right balance between these two conflicting goals. Given the purpose of IAMs—to educate rather than predict—my personal preference is to err on the side of transparency and simplicity, often at the expense of precision and accuracy. That is the essence of abstraction—to focus in on the core operation of a system and learn from the resulting simplified model. The authors of the integrated assessment chapter recognize this (Weyant and others 1996, 392), yet the sense of the chapter is that more detail is a virtue. In the discussion of full-scale IAMs (377–79), the underlying message is that a full-scale IAM is desirable and that it should include all detail possibly relevant to climate change. For instance, the authors suggest that IAMs include local weather and air quality (379). Despite the fact that local air quality has a tie to climate change, this issue clearly should not cloud an IAM, except at the simplest level. Far too many forces are involved in determining local air quality in the many urban areas around the world to hope to be able to represent them realistically in a global model. The authors also suggest that ecosystems be modeled in all their complexity, including the allocation of water resources among competing ends (379).

In my mind, the modeling community must grapple with the tension between complexity and transparency. The ultimate test of transparency is whether a model's structure can be communicated to peer groups. Perhaps some effort should be made to improve methods of communicating the structure of models. This might include a requirement that for an IAM to be considered seriously, its structure must be communicated to the community in an effective way—through a short paper for instance, not a large report. For a model to be accepted, documentation is a necessary but not sufficient condition; a bulky volume documenting a model does make the model structure readily accessible to the modeling community. On the other hand, transparency can obviously be taken too far. To foster realism in IAMs, perhaps standard simulations could be presented, showing how the IAM performs in a standardized context.

ACCOMPLISHMENTS OF INTEGRATED ASSESSMENT

What have we learned from integrated assessment to date? This question is valid though difficult to answer. Chapter 10 lists a variety of results in its section 10.5 (see my review in the earlier section, What Is Integrated Assessment?). At the risk of being overly critical, it appears to me that some IAM accomplishments have been inflated while others have been omitted. Furthermore, the chapter focuses too much on relating conclusions from specific models (such as the multiple references to specific results from the IMAGE model or the discussion of the PAGE model outputs on page 389) rather than the general conclusions that have emerged from the IA community. Models generate a plethora of results, most of which do not advance our general level of knowledge. It is important to distinguish between specific model results and general conclusions that have emerged from integrated assessment modeling. In the discussion below I adopt the taxonomy of the integrated assessment chapter, categorizing models as to whether they are policy evaluation or policy optimization and, further, if they are stochastic.

Policy Evaluation Models

Consider the discussion in the integrated assessment chapter of results from policy evaluation models (summarized in the section of this paper, What Is Integrated Assessment?). It would appear that nearly all of the results listed (a) can be obtained from nonintegrated disciplinary models or (b) are of questionable usefulness. For instance, one result cited is that regional demands for land can serve as a surrogate for measuring local land-cover changes. This question should be answered by micro-level analyses and then serve as an input to an IAM.

Chapter 10 also discusses the sulphate-aerosol debate and the "result" from IAMs that increased forcing due to reductions in sulfur emissions over the next decade could offset the effect of reducing CO_2 emissions. This is a valid conclusion but it is debatable whether one needs an IAM to reach the conclusion. Another result in the integrated assessment chapter is that IPCC scenarios lead to climate-change rates that are higher than those specified by an expert group. This is hardly a general conclusion of IAMs. A last result listed is the effect of a delayed control of greenhouse gases. The result as stated in the integrated assessment chapter is that delaying emissions control by ten years will have a very modest effect on global mean temperature; however, such a delay would necessitate increased reductions of greenhouse gases later to meet concentration goals. Therefore timing is important. This is potentially an important result from the policy evaluation IAMs, although the results of Schlesinger and Jiang (1991) and Hammitt and others (1992) have focused more on this particular question than the cited Rotmans (1990).

Policy Optimization Models

The discussion of results from policy optimization models in section 10.5.2 of the integrated assessment chapter lists three basic results of this type of model. First, optimal emission control rates are small (9% for DICE), supported by a correspondingly small carbon tax ($5/t for DICE). Second, the discount rate matters: lower discount rates can justify control levels as high as 50%. Third, aggressive control now is more costly in the long run than moderate control now followed by more aggressive control later. A fourth conclusion is that adjustment costs may justify a gradual increase to high levels of control, implying that there should be more control now. While I do not disagree with these conclusions, I would characterize what has been learned from policy optimization IAMs slightly differently.

Rather than characterize the DICE result of 9% control as what we have learned, I would suggest that DICE and related analyses tell us that the consensus regarding the magnitude of damages does not justify more than a small amount of current control of greenhouse gases. This result is very significant. We learn that the standard list of effects of climate change do not appear to be as significant as the costs of emission control. Yet the intuition of many has been that the problem justifies significant control. This has led a number of authors to "search" for the climate damage that is perhaps missing from the DICE analysis—to search for the logical basis for taking significant action.

The second result (that the discount rate matters) is not surprising although the quantitative result (that a wide range of control levels can be justified by the appropriate choice of discount rate) is a new result for climate change. This important result has led a number of researchers to try to narrow the range of outcomes. Chapter 4 of the IPCC Working Group III report is concerned with the discount rate (IPCC 1996). So the "result" of the IAM focuses research on a particularly important component of the climate debate. However, the discount rate has long been known to be important for long-run policy decisions.[10]

Another result cited in the report regards the timing of emissions control. The integrated assessment chapter correctly points out that integrated assessment models can help identify emissions paths that minimize the costs of meeting a prespecified concentration level. This is a variant of the "timing" question. The issue is this: when does sharp control need to be initiated? Or, put another way, what is the loss from deferring control for a decade or two? Although not stated in the integrated assessment chapter, the general consensus emerging from IAMs is that there is little to lose from a modest deferral of control. One of the earliest papers to reach this conclusion was Schlesinger and Jiang (1991). More recently, Richels and Edmonds (1995) and Wigley and others (1996) have shown that the IPCC targets can be met by deferring control for a few decades and then taking a more aggressive control path.

Similar to the integrated assessment community's response to the Nordhaus DICE result of low control levels, this result on the low cost of deferral has prompted a closer examination of why action today might be preferable to waiting. Three reasons for taking early action have been explored in the literature. One relates to the adjustment costs of changing control levels rapidly. If it is costly to change emissions quickly, then society should smooth the transition to higher control levels. In the economics literature, this is known as adjustment costs—the cost of adjusting the capital stock. A second reason for taking action now is the influence of experience on control costs ("learning by doing"). If learning by doing is significant, then controls today have two benefits: emissions are reduced and knowledge is generated. Learning by doing might be represented by the costs of control being a function of cumulative experience with control capital. With more and more experience with techniques for reducing GHG emissions, the costs of reducing emissions will tend to decline. Action might take the form of investing in research and development. Although not emissions control, accelerated investment in R&D can be justified on the basis of reducing net costs in the future.

Uncertainty and Learning

Another set of results is associated with uncertainty and information acquisition. The integrated assessment chapter presents some general information on treating uncertainty within IAMs. For the most part, however, the discussion is general.[11] The most substantive result cited in that chapter comes from Nordhaus's DICE model: the level of control and the associated carbon tax rise substantially when parameters in the DICE model are treated as uncertain; the objective is to maximize the expected value of a utility stream.[12] This result has not been reported by others (to

my knowledge) so it may be premature to say that this is a general conclusion of IAMs.

In contrast to the suggestion in the integrated assessment chapter, I
believe considerable work has in fact been done on the questions of
uncertainty and learning (information acquisition), and some general
consensus has emerged. At least three different models have examined
the question of uncertainty and the resolution of uncertainty (Manne and
Richels 1992; Kolstad 1993; Nordhaus 1994). The issue is what effect does
the rate of information acquisition have on today's (1996) emissions control rate? These authors have looked at the question of obtaining perfect
information at different points in time (1995, 2005, 2015, 2025) or the slow
resolution of uncertainty over the 1995–2025 period. They generally conclude that the rate of learning has no effect on the optimal level of control
prior to resolution of uncertainty.[13] In other words, the fact that we may
know in twenty years the true nature of the climate-change problem has
very little effect on the optimal level of emissions control that we should
undertake today.

Although it is a little difficult to fully explain this result, the reason
appears to lie in the relatively slow evolution of the climate compared to
the rate of resolution of uncertainty (see Kolstad 1996). Any mistakes we
might make now, by overemitting, can be undone in the future, after
knowledge is acquired, by underemitting. In essence, we can operate on
the basis of expected value, which means that the presence of learning
can have no effect, ex ante.

General Progress in Integrated Assessment

As a final accomplishment, introducing uncertainty into integrated
assessment has brought improvements in IA methodology. Leaving aside
the work of Nordhaus in the 1970s and early 1980s, the two initial integrated, policy optimization models/analyses were Manne and Richels's
(1990) analysis of the effects on the United States of a carbon tax and
Nordhaus's (1991) steady-state model of climate change. The DICE model
was a substantial step forward from the steady-state model; and Global
2100, MERGE, and CETA were major steps forward from the earlier work
by Manne and Richels. Structurally, these models remain the basis for
most policy optimization models in use today.

In a review of key policy issues in integrated assessment of climate
change for the U.S. Department of Energy, Turner and O'Hara (1996) concluded that one of the five basic issues facing IA was the need to better
represent uncertainty. Two of the biggest problems in representing uncertainty, and the closely related issues of information and learning, are
computational and methodological. It is very difficult to represent uncer-

tainty, let alone information and learning, within integrated assessment models. The Manne and Richels (1992) and Nordhaus (1994) books made major advances by incorporating stochastic elements and learning (though simple). Others have advanced the structure of stochastic control models even further (for example, Kelly and Kolstad, forthcoming).

FUTURE DIRECTIONS FOR
INTEGRATED ASSESSMENT MODELS

Where can we expect integrated assessment models to go in the future? This is anyone's guess, but for the most part the research agenda has already been set or at least suggested.

CHALLENGES AS SEEN BY THE IPCC

Chapter 10 of the Working Group III report suggests that five big challenges face integrated assessment modeling: (1) valuing and representing impacts, (2) dealing with catastrophes, (3) integrating and managing large models, (4) representing developing countries, and (5) improving the relevance of IAMs to policymakers. While these are worthy goals, in general I do not see them as fundamental challenges facing IA over the coming years. The major exception is catastrophes. Not only do we need a better grasp of what catastrophes might befall us, we need to develop better methods for representing catastrophes within integrated assessment models.

Better measures of the impacts of climate change are important to IAMs but must come out of disciplinary studies of ecological or health effects. These can then be used directly in IA. In other words, the problem of valuation in integrated assessment is not with the models but with the underlying basic sciences such as biology, ecology, and economics.

Integrating and managing large models runs counter to one of the messages of this paper that small is usually better. Of course model management is potentially useful. However, the problem of how to manage large models has to be a low-priority challenge.

Clearly, we need to better represent developing countries, but it is not clear that this is a fundamental challenge—it is simply a plea for more coverage of the developing world. Many of the difficult aspects of climate change will be in the developing world—both control of emissions and damage. Furthermore, equity considerations are compounded in the developing world, where, for instance, we are faced with trading off damages to very poor people in Bangladesh with control costs in developed countries.

Improving the relevance of IAMs to policymakers is always important. For the most part, IAMs would seem to be meeting the challenge well. Remembering that the purpose of IAMs is to advance knowledge in the area of climate-change policy and not to advise specific policymakers, the IA community has risen to this challenge admirably.

Other Challenges for IAM

One issue concerns metrics of climate change other than global average temperature. Modelers are aware of the fact that regional differences in temperature, differences in precipitation, frequency of extreme weather events, and other such measures of the climate are important yet largely missing from IAMs. The problem is characterizing damage from these other dimensions of climate change as well as representing the link between global levels of greenhouse gases and these measures of climate. These are not problems of integrated assessment but of the disciplines underlying integrated assessment. Until these "disciplinary" gaps are filled, it is unlikely much progress will be made on this important front.

A related question concerns the rate of change of climate. Clearly the rate of change in temperature (as well as other measures of climate change) has as important a role in damages and the absolute level of climate. The problem here is in characterizing the aggregate damages that result from the rate of change. Some integrated assessment models (for example, Peck and Teisberg 1994) have included the rate of change of temperature, but for the most part these exercises are merely illustrative. Once again, the problem is fundamentally a problem of the underlying disciplines. Not only are rates of change in climate difficult to "predict," but the damages that result from different rates of change are difficult to characterize. Those problems are probably more significant than the problems of including rates of change in integrated assessment models.

Regional variations in climate are increasingly believed to be better indicators of potential damage than averages over broad geographic regions (such as the world). If the global temperature changes by 1°C while the temperature of a region such as the U.S. Midwest changes by significantly more, then damage levels based on the global average will understate damage. The question remains open: is this a fundamental issue for integrated assessment or does it rest primarily in the arena of GCMs (which must be relied upon for forecasts of regional climate change)?

The area of adaptation has remained largely untouched in integrated assessment models (with a few exceptions, such as Hope 1993). By adaptation, I mean adjustments that economic agents make to changes in climate. This includes dike building to ward off elevated sea levels and the development of new hybrid seeds better able to cope with changed climate conditions. Early models of damage assumed no adaptation, which led to high estimates of damage from climate change. To use agriculture as an example, with no adaptation a farmer would continue planting the same crop even though the climate had changed significantly. Later models of damage assume perfect and instantaneous adaptation, which tends to reduce damages significantly. Once again, to use the agricultural example, the model assumes the farmer adjusts practices instantaneously as the climate changes. The truth probably lies between these two extremes, particularly if agents are imperfectly informed about climate change and if climate follows a stochastic process from which it is hard to determine when change has occurred. Since perfect foresight tends to result in decisions with the best outcomes, the role of imperfect information in adaptation may be where we find significant damages from climate change. Furthermore, the response of the R&D sector to changes in climate is very difficult to represent but likely to be important. How quickly will seed companies develop new hybrids to counteract climate change? These issues are important for integrated assessment but largely ignored.

Expanding the menu of policy options is another fruitful direction for integrated assessment. Controlling climate change today involves a significant investment on the part of the present generation that benefits future generations. As Thomas Schelling has pointed out, other investments we may make today (such as accelerating development of the Third World) may have much greater payoff for the future. Population control is another example of a policy with tremendous implications for climate change (see Kelly and Kolstad 1996). Clearly the menu of policy alternatives needs expanding, and integrated assessment is the avenue through which this menu expansion is most likely to occur.

Catastrophes also relate to uncertainty and learning. While some work has been done in this area (for example, Peck and Teisberg 1995), little consensus has emerged on the implications of large-damage/low-probability events for climate-change policy. Furthermore, integrated assessment has had little to say on this issue. This is a natural area for analysis, and I expect progress to be made on this front. There are really two problems here. First, disciplinary research must identify what catastrophes might occur and what their probabilities are. Currently there is little appreciation for what kinds of cataclysmic events might occur. Identifying such events is not in the domain of integrated assessment. After catastrophes have been characterized, IA can incorporate such extreme events into models.

The area of noncooperative analysis of the evolution of climate policy and economies in general has received little attention in either Chapter 10 or the integrated assessment community—although a few models have been developed which do not assume cooperation in reducing greenhouse gases (for example, Nordhaus and Yang 1996). One thing is clear to many: a global all-inclusive treaty involving tradeable permits or a uniform emissions fee will not likely materialize. Second-best alternatives are the best we can hope for. Considering the problems of initiating and enforcing agreements with large numbers of parties, actions and treaties involving small numbers of countries must be considered a possibility. Integrated assessment models must be able to analyze how such mixed cooperative-noncooperative structures might operate and evolve. Some of the simpler policy optimization models are reasonably well suited for this type of analysis. To my knowledge, no model endogenizes the formation of voluntary coalitions to pursue reductions in greenhouse gas emissions.

A last, though by no means least, important future direction is in representing technological change and R&D. Ideally, public R&D should be considered a policy option to be adjusted to the optimal level. Private R&D responses to incentives for the control of greenhouse emissions are also undoubtedly significant. This has been mentioned as an important area of current research from which a few results have emerged. I expect this area to bear considerably more fruit over the coming decade.

CONCLUSIONS

Considering the relatively short history of integrated assessment of climate (less than a decade), a surprising amount of knowledge has emerged. Probably the most striking result is that our current understanding of the damage of climate change does not justify more than modest emissions control (though that still would be more than is being pursued currently). Other results, though not as fundamental, have also emerged from integrated assessment. These include the sensitivity of policy to our treatment of future generations (the discount rate question) and the finding that information acquisition makes little difference to current-period levels of emissions control.

In reviewing the fundamental knowledge of climate-change policy that has emerged from IA, strikingly, nearly all of the results have come from the so-called policy optimization models, the top-down economy-climate models. Virtually no new basic understanding appears to have emerged from the policy evaluation models (with the possible exception of some insights on cooling from sulfates). Two possible alternative conclusions can be drawn from this. First, the list of accomplishments from integrated assessment is incomplete and skewed toward the policy optimization models. Second, the more complex policy evaluation models are not as useful in the integrated assessment process. This issue deserves more consideration and debate in the integrated assessment community.

ACKNOWLEDGMENTS

This work was sponsored in part by U.S. Department of Energy grant FE03–94ER61944. Comments from Jae Edmonds, Brian Flannery, Erik Haites, Granger Morgan, John Reilly, Rich Richels, Dick Schmalensee, and John Weyant have been appreciated. I am particularly grateful to Bill Nordhaus for detailed comments on several versions of the manuscript.

ENDNOTES

1. This chapter (Weyant and others 1996) is referred to here as "the integrated assessment chapter."

2. An exception to this generalization is integrated assessment models developed to support a specific policy. For instance, if the Environmental Protection Agency is statutorily required to determine regulations on emissions of specific air pollutants, it will often use the results of a detailed model to generate insights on regulations. An example of this is discussed later in this paper in the context of the U.S. regulations on sulfur emissions from coal-fired power plants.

3. A nice discussion of possible policy goals of integrated assessment can be found in Dowlatabadi and Morgan (1993).

4. Refer to a series of papers culminating in Nordhaus (1994).

5. For instance, on page 18 of the integrated assessment chapter, "several of the key models omit direct modeling of economic activity and rely on exogenous greenhouse gas emission trajectories."

6. The Limits to Growth study (Meadows and others 1972) and the analysis of the supersonic transport (SST) are mentioned in the integrated assessment chapter but not much of the other work from that era.

7. See Ackerman and Hassler (1981) and Kolstad (1990). At issue was the extent to which flue gas desulphurization (FGD) would be required on new coal-fired power plants.

8. See Tietenberg (1996) for a review of some of these studies.

9. The only possible exception to this statement is that large, well-supported models can be useful, but the level of support necessary is very substantial. Support must include complete documentation, rigorous data development, validation, reliability, broad communication of model structure to peers, and acceptance of that structure by peers. This level of support has yet to materialize in the climate arena, in part because the climate problem is so large and diffuse.

10. See the volume from the early 1980s on discounting in energy policy (Lind and others 1982), as well as the debate in the 1960s on water resources projects (for example, dams).

11. See, for instance, the discussion on pages 389–90 (Weyant and others 1996) of different approaches for dealing with uncertainty.

12. This is a little difficult to deduce directly from the Nordhaus (1994) results. His Table 8.2 provides a *certainty equivalent* (CE) case where the model parameters are set at their expected values. But he does not provide a case where the expected value of utility is maximized, except over the five aggregate states-of-the-world. It is not clear whether the expected values of all the parameters over the five states-of-the-world are the same as those used in the certainty equivalent case.

13. This is qualified somewhat when long-lived abatement-specific capital investment is considered. In this case, information acquisition may tend to depress today's control level (Kolstad 1994; Grubb and others 1995).

REFERENCES

Ackerman, B. and W. Hassler. 1981. *Clean Coal, Dirty Air*. New Haven, Connecticut: Yale University Press.

Alcamo, J., R. Shaw, and L. Hordijk, eds. 1990. *The RAINS Model of Acidification: Science and Strategies in Europe*. Dordrecht, Netherlands: Kluwer.

Atkinson, S. and D. Lewis. 1974. A Cost-Effectiveness Analysis of Alternative Air Quality Control Strategies. *Journal of Environmental Economics and Management* 1:237–50.

Atkinson, S. and T. Tietenberg. 1982. The Empirical Properties of Two Classes of Designs for Transferrable Discharge Permit Markets. *Journal of Environmental Economics and Management* 9:101–21.

Dowlatabadi, H. and M. G. Morgan. 1993. Integrated Assessment of Climate Change. *Science* 259(5103):1813.

Grubb, M., M. Duong, and T. Chapius. 1995. The Economics of Changing Course. *Energy Policy* 23:417–32.

Hammitt, J. K., R. J. Lempert, and M. E. Schlesinger. 1992. A Sequential-Decision Strategy for Abating Climate Change. *Nature* 357:315–18.

Hope, C. 1993. Policy Analysis of the Greenhouse Effect. *Energy Policy* 21:27–38.

IPCC (Intergovernmental Panel on Climate Change). 1996. *Climate Change 1995: Economic and Social Dimensions of Climate Change. The Contribution of Working Group III to the Second Assessment Report of the Intergovernmental Panel on Climate Change*. Edited by J. P. Bruce, H. Lee, and E. F. Haites. Cambridge: Cambridge University Press.

Kelly, D. and C. Kolstad. 1996. *Malthus and Climate Change: Betting on a Stable Population*. Department of Economics Working Paper 9-96R. September. University of California-Santa Barbara.

————. Forthcoming. Bayesian Learning, Growth and Pollution. *Journal of Economics Dynamics and Control.*

Kolstad, C. 1986. Empirical Properties of Economic Incentives and Command-and-Control Regulations for Air Pollution Control. *Land Economics* 62:250–68.

————. 1990. Acid Deposition Regulation and the U.S. Coal Industry. *Energy Policy* 18:845–52.

————. 1993. Looking vs. Leaping: The Timing of CO_2 Control in the Face of Uncertainty and Learning. In *Costs, Impacts and Benefits of CO_2 Mitigation*, edited by Y. Kaya, N. Nakicenovic, W. D. Nordhaus, and F. L. Toth. June. Laxenburg, Austria: IIASA .

————. 1994. George Bush vs. Al Gore: Irreversibilities in Greenhouse Gas Accumulation and Emissions Control Investment. *Energy Policy* 22:771–78.

————. 1996. Learning and Stock Effects in Environmental Regulation: The Case of Greenhouse Gas Emissions. *Journal of Environmental Economics and Management* 31:1–18.

Lind, R. C., K. A. Arrow, G. R. Corey, eds. 1982. *Discounting for Time and Risk in Energy Policy.* Baltimore: Johns Hopkins University Press.

Manne, A. S. and R. G. Richels. 1990. CO_2 Emission Limits: An Economic Cost Analysis for the U.S.A. *Energy Journal* 12:87–107.

————. 1992. *Buying Greenhouse Insurance: The Economic Costs of CO_2 Emission Limits.* Cambridge, Massachusets: MIT Press.

Manne, A. S., R. Mendelsohn, and R. G. Richels. 1993. MERGE: A Model for Evaluating Regional and Global Effects of GHG Reduction Policies. *Energy Policy* 23:17–34.

Meadows, D. and others 1972. *The Limits to Growth.* New York: Universe Books.

Nordhaus, W. D. 1991. To Slow or Not to Slow: The Economics of the Greenhouse Effect. *Economic Journal* 101:920–37.

————. 1994. *Managing the Global Commons.* Cambridge, Massachusets: MIT Press.

Nordhaus, W. D. and Z. Yang. 1996. A Regional Dynamic General-Equilibrium Model of Alternative Climate-Change Strategies. *American Economic Review* 86:741–65.

Peck, S. C. and T. J. Teisberg. 1993. Global Warming Uncertainties and the Value of Information: An Analysis Using CETA. *Resource and Energy Economics* 15.

————. 1994. Optimal Carbon Emissions Trajectories When Damages Depend on the Rate or Level of Warming. *Climatic Change* 30:289–314.

————. 1995. Optimal CO_2 Control Policy with Stochastic Losses from Temperature Rise. *Climatic Change* 31:19–34.

Richels, R. and J. Edmonds. 1995. The Costs of Stabilizing Atmospheric CO_2 Concentrations. *Energy Policy* 23:373–78.

Rotmans, J. 1990. *IMAGE: An Integrated Model to Assess the Greenhouse Effect.* Dordrecht: Kluwer Academic Publishers.

Rubin, Edward S. 1991. Benefit-Cost Implications of Acid Rain Controls: An Evaluation of the NAPAP Integrated Assessment. *Journal of Air and Waste Management Association* 41:914–21.

Rubin, E. S., L. B. Lave, and M. G. Morgan. 1992. Keeping Climate Research Relevant. *Issues in Science and Technology* 8(2):47–55.

Schlesinger, M. E. and X. Jiang. 1991. Revised Projection of Future Greenhouse Warming. *Nature* 350:219–21.

Tietenberg, T. 1996. *Environmental and Natural Resource Economics*, 4th ed. New York: HarperCollins.

Turner, R. S. and F. M. O'Hara. 1996. *Integrated Assessment of Climate Change: Characterizing Key Policy Issues*. February. Center for Global Environmental Studies. Oak Ridge, Tennesee: Oak Ridge National Laboratory.

Weyant, John, O. Davidson, H. Dowlatabadi, J. Edmonds, M. Grubb, E. A. Parson, R. Richels, J. Rotmans, P. R. Shukla, R. S. J. Tol, W. Cline, and S. Fankhauser. 1996. Integrated Assessment of Climate Change: An Overview and Comparison of Approaches and Results. Chapter 10 in IPCC 1996.

Wigley, T., R. Richels, and J. Edmonds. 1996. Economic and Environmental Choices in the Stabilization of Atmospheric CO_2 Concentrations. *Nature* 379:240–43.

Comments

Integrated Assessment Modeling of Climate Change

John P. Weyant

This note is a short response to Charles Kolstad's review of Chapter 10, Integrated Assessment of Climate Change, from the 1996 report of IPCC Working Group III. The review raises many interesting and provocative issues, some of which I address here. Dr. Jae Edmonds, another lead author of the chapter, addresses others in his response. I will focus here on but three of the many issues addressed in the Kolstad review:

1. The advantages of simple models versus those of complex ones.
2. The purposes of integrated assessment.
3. Lessons from integrated assessment in other areas.

For the most part I accept and expand upon the Kolstad critiques on issues (2) and (3) and offer some rebuttal on (1).

SIMPLE VERSUS COMPLEX MODELS

There are a number of dimensions to the debate about whether simple or more complex models are most useful to those involved in policy development and analysis. Obviously, simple models are more transparent to users than more detailed models. On the other hand, the details included

JOHN P. WEYANT is Professor of Engineering-Economic Systems and Operations Research, and director of the Energy Modeling Forum at Stanford University.

in the more complex models may be more familiar to users than the aggregate, and sometimes arcane, inputs and parameters of the simpler models. For example, individual technology projections may be more meaningful to users than aggregate demand and substitution elasticities. In addition, size need not always add complexity to a model, if the size comes from disaggregation of data or relationships by region, sector, household type, industry, and so on. On the other hand, some users (and analysts) ascribe greater certainty to the inclusion of greater detail in the model. As has been demonstrated in a number of areas, this need not be the case. In fact, one problem with large, complex models is that they can not easily be used to do uncertainty analyses—which can be handled with the simpler type of models.

In the specific case of climate change, two additional questions arise regarding the usability of results from integrated assessment models. First, are aggregate damage functions, which relate global GDP losses to changes in mean global temperature, credible to policymakers? They provide a useful benchmark for the preliminary design of policy responses. But at some point policymakers will want to know if an increment of GDP loss is the result of fewer recreational amenities (fishing, hiking, beach days) in Southern California or more deaths from Monsoon-driven flooding in Asia or climate-driven increases in malaria incidence in Africa.

The second potential problem with using only simple models in policy analysis is that they may not explicitly include the policy levers preferred by policymakers. For example, the simple model may use a carbon tax as a proxy for measures designed to efficiently reduce carbon emissions while the policymaker may want to consider energy-use efficiency standards or R&D options.

The solution to the problem of choosing a level of complexity for a particular analysis may lie in the development of hierarchical model architectures, in which details can be maintained near the point of policy implementation, and less detail can be used (while maintaining internal consistency) on areas of the economy not directly affected or relevant. The more aggregate representations could in themselves be "reduced-form" representations of more complex models used in other analyses that focus on them more explicitly. With this strategy we can also allow for the formal or informal integration of simple, complex, and judgmental models (including results from expert-opinion surveys) within a single analysis. Indeed, the integration between the various components included in the analytical framework could be completely automated or the user could be given the option to intervene at selected interface points.

PURPOSES OF INTEGRATED ASSESSMENT MODELING

As noted by Kolstad, Chapter 10 overstates (probably unintentionally) the likely direct role of integrated assessment modeling in policy development. Rather than using models directly in the policy development process, analysts will more likely use them to learn about the implications of various policy options. Then, the analysts will use these insights to influence the thinking of policymakers, and ultimately the public. So, initially at least, the purpose of integrated assessment is to gain insights, not numbers. Significantly, integrated assessment can be used to identify smart and dumb policies and to identify previously unrecognized interactions and feedbacks in the system being analyzed.

At some point, however, policymakers will want numbers to help guide the actual implementation of whatever policies look most desirable. Maybe they should not be so interested in these details. But the need to trade off climate-change policies against policies on other pressing issues may make this search for numbers unavoidable. And the models may provide the only credible recourse for producing the numbers that are required.

However, these numbers ought never be used without some degree of sensitivity and uncertainty analysis to give the policy process a feel for how robust policies based on their use might be.

LESSONS FROM IA OUTSIDE OF CLIMATE CHANGE

During the course of putting together Chapter 10, we made a strategic decision to stick primarily to a review of IA of climate change and closely related areas. This decision was based primarily on time and resource constraints, and, as cogently argued by Kolstad, was probably a mistake. Insights about the process of integrated assessment are to be found in the vast literature on energy, environmental, and other issues; and a focus on insights from these studies, rather than a comprehensive review, would have been a very useful addition to the chapter.

Kolstad concludes from his own review of this literature that simple IA models have been more useful than complex IA models. While agreeing with that assessment, I would draw a somewhat weaker implication from the evidence—that complexity has generally not been added strategically to IA models focused on particular policy issues. Part of the underlying cause for this result has been that decision relevance has often been subjugated to process understanding. In other words, science has been pursued in and of itself without considering the likely policy relevance of that science. In my view, this approach is antithetical to the reason IA models are constructed.

Thus, I support Kolstad's conclusion that the experience with IA to date has generally supported the advantage of simple over complex IA models. However, I believe this situation can and should improve as the IA of climate change progresses. Better theory, more experienced modelers, more powerful computers, more user-friendly software, and more sophisticated model users have set the stage for more strategic and useful addition of complexity to integrated assessment models of climate change.

Comments

Integrated Assessment Modeling of Climate Change

Jae Edmonds

In his review of the IPCC Working Group III Chapter 10 (Weyant and others 1996), Charles Kolstad is generally supportive of the field of integrated assessment (IA), being an accomplished practitioner. (See for example, Kelly and Kolstad 1996, or Kolstad 1992, 1994.) Beyond that, Kolstad is generally favorably disposed to the treatment of IA by Weyant and others (1996), a view to which I also subscribe. My views, however, are clearly colored by my own participation as a lead author on the chapter. Kolstad's review goes beyond simply providing a set of criticisms of Weyant and others, to take, as Kolstad puts it, "a fresh look at the subject that has become known as integrated assessment modeling, and to offer a few observations that can perhaps focus thinking about future agendas."

IDENTIFYING IA IN ITS NATURAL HABITAT

The first substantive question Kolstad addresses is how do we know IA when we see it? What distinguishes IA models (IAM) and modeling from the vast body of disciplinary research? And why would rational researchers practice the IA arts? Kolstad contends that "the primary role of integrated assessment is to analyze complex questions that can only be

JAE A. EDMONDS is chief scientist in the Global Change Group at Pacific Northwest National Laboratory.

answered by integrating multiple disciplinary aspects of the climate problem." He goes on to argue that a model must have a human-activity component to be an IAM. This is a significantly narrower definition than that which Weyant and others construct and is also narrower than the one he uses to define the purpose of IA in general. He argues, for example, that a model that combines an exogenously specified emissions trajectory with a general circulation model should not be considered an IAM. This narrower definition of IAMs places economics in a central role and is in keeping with the spirit of the development of IAMs in the United States.

But this narrower definition misses the main thrust of IAM development in Europe and, to some extent, Asia. The proposed definition implies, for example, that the IMAGE 2.0 model (Alcamo 1994) is not an IAM because it does not include human economic activity in a meaningful way. But IMAGE 2.0 is clearly not a disciplinary model and is widely recognized as a leading IAM in Europe. Without question the IMAGE 2.0 model is different in its structure and motivational origins, but it incorporates information from a diverse set of sources and has had an important influence on the formulation of an understanding of the climate issue. The IMAGE 2.0 model focuses on the natural-science components of the problem. It integrates ecosystem, land-use, and atmospheric process models within its system architecture. These components shed light back on climate-change decisionmaking, but from a different perspective than those taken by models such as DICE (Nordhaus 1994), MERGE (Manne and others 1995), ICAM (Dowlatabadi and Morgan 1993), or MiniCAM (Edmonds and others 1996). Where the cost-benefit and decision-analytic framing of the problem are paramount in the latter group of models, the "precautionary principle" is dominant in IMAGE 2.0. The IMAGE 2.0 model is better suited to provide information within the "precautionary principle" paradigm than models such as DICE, MERGE, MiniCAM, or EPPA. I find the narrower definition of IAMs too restrictive. It unnecessarily excludes models that are clearly interdisciplinary, and it is the interdisciplinary character of the research that defines IA, not the joining of social- and natural-science components.

I personally define IA as follows: Integrated assessment for climate change assembles knowledge from a diverse set of sources, relevant to one or more aspects of the climate-change issue, for the purpose of gaining insights that would not otherwise be available from traditional, disciplinary research.

In the final analysis, however, the case for or against IAMs cannot depend critically on the precise position of technical boundaries defining the activity. The field's contribution will depend primarily on the contributions of sustained research activities which define themselves as IAMs.

NEVER SAY "NEVER"

With some significant experience in the area of IAM already available, Kolstad proposes two generalizations for application to the IAM activity. First, "complex models are of limited usefulness, particularly if their structure is nontransparent." Second, "forecasts per se are less interesting than insights into the problem."

All Models Great and Small

Big, complex (that is, "great") models are limited in several ways. First, they are expensive and time consuming to build. Great models do not come into existence over night. They take years to assemble and validate. And when completed, their workings may be difficult to explain outside the technical community. Second, they can address only a limited type of question. Their complex structure will have been designed to address a specific set of questions. If the policy debate suddenly shifts to other questions, it may be exceedingly difficult to refocus the model for one of two reasons. Either the new question is naturally suited to smaller models (due to its inherent computational burdens), or the core detail of the large-scale model is rendered irrelevant by the shifting policy debate. Third, great models cannot adequately address whole classes of problems, such as optimization or uncertainty analysis, by virtue of their computational costs.

To the extent that they are built within an "official" context, great models, and particularly great models which embody nontrivial economic elements, must carry another burden. They become the only game in town. While they are potentially the "stars" of the show, they may also become the "villains." Their reference case becomes a forecast rather than a base against which to assess deviations in research experiments. For researchers this can be an uncomfortable and confusing position. Furthermore, in a world in which political perspectives differ, constituencies for whom the "official model" is not producing comforting results have an incentive to attack the model. The political process needs room in which to work. To the extent that only one optimal policy is being generated by an "official" IAM, at least some participants in the political process have been backed into a corner and may conclude that the model has to go. The model can have a very difficult time surviving. This was the fate of integrated assessment modeling undertaken by the National Acid Precipitation Assessment Program (NAPAP). NAPAP was constantly frustrated in its efforts to develop a large-scale, official IAM, and the tool that was finally developed was of reduced scale and never influenced policy discussions.

The European community has dealt with the problem in two ways. First, there is a greater willingness in Europe to defer to experts. The skepticism that pervades the United States on such matters is less evident on the eastern shore of the North Atlantic. The great model is simply an expert tool. Furthermore, as noted earlier, great models focus primarily on knitting together natural-science issues, leaving aside social science and especially economics. The RAINS model (Alcamo and others 1990) was able to play an important role in the European sulfur-negotiations process. But it did so by abdicating the responsibility for independently developing emissions trajectories. The RAINS model simply adopted emissions inputs brought to them by negotiators, who in turn took the acid deposition estimates given by the RAINS model as truth.

The present suite of IAMs are of a third character. They are multiple and diverse. They vary in size, including small, medium, and great models. Integrated assessment models have been developed employing a wide variety of approaches, including different levels of geographic and economic aggregation, breadth of greenhouse-related gases, and sophistication of atmospheric and climate processes. The models' institutional and geographic origins are also highly varied. Models have been developed in many OECD countries in Asia, North America, and Europe. No model plays the role of "the official" model for any nation or international organization. The current set of integrated assessment models can potentially play an important role in the political process because their diversity of output spans a wide spectrum of results, enabling the political process to pick and choose in building their political cases. However, the set of results are importantly limited by the characteristically common process representations embedded in all the IAMs.

The new IA environment provides a place in which large models can participate in the scientific and political discussions in a way that they could not, for example, under NAPAP. IAMs have been freed from the role of "official" keepers of all truth and can now provide targeted insights to the climate issue. To date the most active great models have been IMAGE 2.0 and AIM (Morita and others 1993). IMAGE's success is due to two factors. First, the IMAGE group was an assembled, seasoned team at the time the IMAGE 2.0 effort began. IMAGE 2.0 was made possible by the experience gained at RIVM in the development of IMAGE 1.0 (Rotmans 1990). IMAGE 1.0 was small and closer in spirit to the current crop of IAMs in the United States. Interestingly, the PNNL GCAM (Edmonds and others 1993) modeling effort follows a similar path, with the MiniCAM providing a test bed for the introduction of new concepts within a small-scale modeling framework, and the PGCAM providing a format for the full, process-level implementation. Second, the scope of IMAGE 2.0 was narrower than that taken by the large-scale U.S.-based models such as MIT (Prinn and others 1996) and PNNL (Edmonds and others 1993). It focused primarily on integrating natural-science issues to the exclusion of social science. AIM has succeeded by focusing its efforts on the Asia-Pacific region.

The Devil's Details. The principal advantage of great models is that they provide process detail. They fill in the picture. Kolstad's second lesson is that "forecasts per se are less interesting than insights into the problem." Certainly one cannot argue that insights are unimportant. Insights are critical. They are the difference between numbers and knowledge. The current suite of predominantly simple IAMs have been particularly powerful in shaping understanding of the climate issue when they reach common, qualitative results.

Kolstad recognizes the tension that exists between the competing modeling goals of detail and realism on the one hand and simplicity and transparency on the other. He comes down on the side of simplicity and transparency. It is hard to oppose simple and transparent models. Their results are easy to explain. They speed sorting through outputs and culling unreasonable findings. They are quick to build and run. But simplicity and transparency are relative terms. One modeler's simplicity is another modeler's complexity. In Edmonds and Reilly (1985) John Reilly and I argued for a concept that we called "minimum modeling." The idea is similar in spirit to that espoused by Kolstad, but is importantly relativistic. The concept of "minimum modeling" argues that we work from the problem to the model, not the other way round. To do so, the model should focus on explaining the critical elements necessary to understand the issue at hand, and no more. But there are a wide variety of issues with which to deal. The present class of simple models—such as MERGE, DICE, ICAM, and MiniCAM—were originally designed to deal with highly aggregated problems such as identification of the optimal level of global aggregate fossil fuel emissions mitigation.

Making the case for insights differs from arguing that numbers don't matter. Insights are derived from numbers. These are computer models after all, and they are ultimately little more than a stream of numbers. Simple models have fewer numbers than complex models. The smaller number of numbers is achieved by aggregating. But ultimately the only difference between simple and great models is in the degree of disaggregation. Where the modeler stops on the scale of disaggregation of any particular problem should, in principle, depend on the problem. It is ultimately a judgment call. The level of detail in MiniCAM is greater in all dimensions than that in DICE. But both are simple models. The MIT IA program uses a two-dimensional climate model. Compared with the state-of-the-art in climate modeling, it too is a simple model, but compared with the zero-dimensional climate models used in many IAMs, the MIT model is complex. All of which raises the question as to the value of time spent in trying to establish lines of distinction between simple and complex models.

Models are only as good as the questions they answer. In many instances the only way we know whether or not an additional level of detail matters is to do the research. The history of important atmospheric-model results is a story of modelers who repeatedly found that the detail they thought to be of third-order importance was in fact of first-order importance.

The desired precision of results will also vary with the state of the policy discussion. In the early stages of the discussion the principal question at hand is, "Does this issue belong on the policy agenda at all?" This is a question that is not entirely analytical. But it includes an analytical component, and that component can weigh in the debate. Here qualitative answers can be extremely important. They can either help the process move on to more pressing matters, or provide guidance as to which areas would benefit from further information that would help clarify the degree of gravity with which to regard the issue. As policy questions become more specific, the degree of precision with which we wish to answer them increases, and additional complexity in the models is essential if the models are to continue to play a role in advanced discussions.

Ultimately the simple models and high level of aggregation must give way. Budgets need to be created and balanced. In that case a billion dollars is a big deal. Specifics of treaties need to be negotiated and a billion tonnes matters. The treatment of fossil fuel carbon only, while a reasonable approach to take in a political triage, will not suffice if the climate problem is found to be something other than a "political no brainer," neither requiring simple, dramatic, and decisive actions, nor amenable to dismissal from the policy agenda.

The fossil-fuel-only assumption harbors the seeds of large policy errors when other greenhouse-related gaseous emissions, such as CO, CH_4, VOCs, N_2O, NO_x, and SO_2, are exerting potentially strong forces on different time scales. Similarly, treating anthropogenic land-use emissions and the terrestrial carbon cycle as if they were in perpetual fortuitous equality may be acceptable for a first pass at the problem, but it will not stand up to closer scrutiny from either the scientific or policy perspectives. Recent research reviewed in IPCC (1996a,b,c) indicates that such simplifications are wholly inadequate. Treating these issues, such as optimal policy and emissions-mitigation patterns to achieve alternative steady-state climate regimes, requires a suite of more sophisticated models. Similarly, the interconnectedness of problems such as climate change, stratospheric ozone, acid deposition, and biodiversity argues for deepening the level of understanding embedded in IAMs.

Policymakers are constantly framing policy responses that eschew the simple policy instrument of taxation in favor of other more sophisticated applications of fiscal, regulatory, information, and research policies. In such instances models need to have sufficient geographic, temporal, and sectoral detail to respond. There is enormous potential for policymakers to choose policies which consume vast quantities of societies' resources and yield little in the way of climate benefits in return. Unless models can address the additional costs associated with more complex policy formulations, they leave open the door for great mischief.

This is not to forget that our ability to foresee the future is limited. We meet limitations at every turn. Climate models agree on the global mean temperature change for a "doubling" of preindustrial CO_2 to less than plus or minus 50%, while economic-emissions models admit to similar degrees of uncertainty in cumulative carbon emissions. Adding detail to address increasingly sophisticated policy questions or to reflect expanded understanding of our world may or may not increase the confidence in which we hold forecasts.

WHAT HAVE IA MODELS DONE FOR US LATELY?

Since IAMs lie at the crossroads of climate-change modeling and assessment, they straddle as a class the full breadth of the climate issue. Yet they are, with some exceptions, composed of reduced form representations of the central components of the problem. For the most part, they assemble information that is being created at the various disciplinary frontiers, so as to uncover and quantify implications and feedbacks that might otherwise go unappreciated. IAMs have produced a variety of results, and as Kolstad correctly points out, model-specific results come and go. Only when models begin to repeatedly obtain the same finding can the community be said to be making a contribution. This is not to say that for some issues and at some times specific modeling teams may not be leading the community into the investigation. But ultimately it is the reproducibility of the experiment that takes a contribution from the domain of interesting result to major contribution.

Four Important Findings

Kolstad sifts through the variety of results presented in Chapter 10 with an eye toward separating those which fall into the category of major contribution from others which are either merely interesting, model-specific results or disciplinary findings. The results that make the cut might be summarized as follows:

1. Optimal emissions-control rates are small and are supported by correspondingly small carbon tax rates.
2. The discount rate matters; lower discount rates can justify control levels as high as 50%.
3. Optimal near-term actions are relatively robust, and IAMs indicate modest near-term emissions mitigation unless severe damage thresholds are anticipated to be relatively near at hand.
4. It is important to incorporate "flexibility" into national and international mitigation programs.

One could quibble that the discovery of the discount rate as an important determinant of policy response in long-term, cost-benefit analysis predates the introduction of IAMs. But the importance of the discount rate in climate-change policy discussions looms so large that the result bears repeating.

The other conclusions are nontrivial and their policy implications are unavoidable. This explains at least in part why Chapter 10 had such difficulties making it through the IPCC approval process. The *Summary for Policy Makers* is a consensus document, approved by the world's governments. Without consensus, there is no approval. And the perceived national interests of governments vary greatly around the world. Some governments were unable to live with the conclusions of the IA chapter. Their nominal objection was that IA was too "new" to support substantive results. But the same could have been said for the long-term aggregate impacts research. That chapter had many problems attributed to it, but research immaturity was not one of them. What made the IA conclusions potentially dangerous was that they left little room for the breadth of political perspective that exists among nations. Such difficulties are also a testament to the importance and power of IAMs.

Timing-related results have been particularly controversial. The paper by Wigley and others (1996) served as a lightning rod for much of this controversy within the IPCC, but similar results have emerged from a variety of other papers dating back as far as Nordhaus (1979). These papers have challenged popular perceptions and while *not* necessarily implying "do nothing," they have raised serious questions as to the most effective near-term strategy for dealing with this long-term problem. IAMs have also shown that the appropriate policy response may involve little mitigation in the near term, with greater emphasis on the preparation of institutions and the development of relevant science and technology for later actions—at least for ceilings of 550 ppmv and higher. The work of Wigley and others indicates that for ceilings higher than 450 ppmv, the appropriate policy mix should initially emphasize the creation of institutions capable of minimizing costs and the development of low-cost, high-efficiency, noncarbon energy technologies capable of large-scale deployment in the middle of the next century. Over time the policy mix should increasingly emphasize policies which control emissions. Over the long term a steady-state concentration of CO_2 can only be maintained with an ever declining net emissions rate, and so ultimately policies must evolve toward prohibition of emissions.

IAMs have also shown that the method of implementing emissions mitigation is just as important as the degree of mitigation undertaken, implying that important near-term policy objectives should be the creation of institutions which minimize emissions-mitigation costs. The costs of achieving a specific emissions-mitigation goal can vary by a factor of five to ten depending on the *flexibility* with which the goal is attained (Richels and others 1996). IAMs have reminded negotiators that it is not just the goal that determines costs but also the manner in which the goal is achieved. Interestingly, the cost of stabilizing the concentration of CO_2 at 550 ppmv with full flexibility (Edmonds and Wise 1996) was lower than the cost to the OECD of reducing emissions 10% in 2010 and holding them constant to 2050 with no flexibility (Wise and Edmonds 1996). IAMs have also begun to map out the relationship between minimum-cost and steady-state CO_2-concentration targets and have found that costs rise sharply as the concentration ceiling declines from 550 ppmv to 450 ppmv.

What is truly remarkable is the degree of impact IAMs have had on framing the policy discussions. They have refocused the debate from emissions to concentrations, the stated objective of the Framework Convention on Climate Change (FCCC). This in turn has focused research and debate on the question of near-term actions which are appropriate in light of the long-term nature of the climate-change problem. IAMs have also focused attention on the issue of implementation. While modelers can act as if a policy extending over the course of the next century can be put in place today and remain unchanged, in the real world the decision-making process is one in which we act, then learn, then act, then learn some more, then act again, and so on for centuries. It is therefore near-term decisions that matter, though they must be conceived in recognition of the long term as it is currently understood.

In a similar vein, IAMs have shown that providing flexible instruments could lower the cost of achieving various mitigation goals by an order of magnitude (Richels and others 1996). Flexibility comes in two flavors: "where" and "when." "Where" flexibility implies employing tools which allow emissions mitigation to go on wherever it is cheapest; "when" flexibility implies allowing emissions mitigation to occur whenever it is cheapest to do so, as long as a cumulative emissions-mitigation constraint is satisfied. Interest in tradeable emissions permits stems at least in part from their potential as a tool for affecting "where" and "when" flexibility.

Sulfur

One result highlighted by Weyant and others (1996), which Kolstad dismisses, is the importance of sulfur emissions. I believe the importance of sulfur in formulating policy should be taken seriously as an IAM result. On a trivial level there is little new in the sulfur story at all. The general properties of sulfate aerosols have long been appreciated by natural scientists. A good summary of the science of aerosols and their relationship to climate change can be found in Wang and others (1985). But it was not until IPCC (1992) that sulfate aerosols were generally believed to explain the discrepancy between GCM output and the observed climate record. Even then natural scientists did not appreciate the implications that this finding might have for policy. This appreciation can be traced to IAM researchers, for example, Edmonds and others (1994) and Ball and Dowlatabadi (1995), who showed that the introduction of sulfur greatly complicates the policy-formulation problem.

Sulfate aerosols exert a powerful negative effect on net radiative forcing, and sulfur emissions are coincident with carbon in fossil fuels, though the relationship between carbon and sulfur occurrences is highly variable. Policies that rapidly reduce carbon emissions can also rapidly reduce sulfur emissions and thereby rapidly unmask the greenhouse effect leading to high, near-term, decadal rates of climate change. Optimization models, which have attempted to introduce sulfur into the IAM framework, respond in the presence of positive discount rates by increasing the rate of carbon and sulfur emissions as a strategy for maximizing welfare. Other IAMs have found that improvements in noncarbon energy-supply technologies, which lead to rapid reductions in fossil fuel carbon emissions, may actually reduce welfare. These complications are real and important.

More recently, the "sulfur result" has become part of the "equivalent CO_2" issue. This is a nontrivial calculation being developed within the context of the negotiation process. The idea is very similar to the concept of a global warming potential (GWP) coefficient. Here we ask what combination of concentrations of CO_2 and other greenhouse gases in the atmosphere yield the same radiative forcing as 550 ppmv of CO_2 alone. To the degree that future background sulfur emissions more than offset the combined effects of other non-CO_2 greenhouse-related gases, real CO_2 emissions are associated with a ceiling higher than 550 ppmv CO_2. And, to the degree that future background sulfur emissions less than offset the combined effects of other non-CO_2 greenhouse-related gases, real CO_2 emissions are associated with a ceiling lower than 550 ppmv CO_2. This is a tricky calculation and can make a 50 to 100 ppmv difference in the "real" CO_2 concentration. It also hinges on the degree to which sulfur (and other greenhouse-related emissions) move with fossil fuel carbon emissions, especially in the downward direction.

The short lifetime of sulfur implies a highly heterogeneous distribution of concentrations and highly localized effects on radiation. It has important implications for determining winners and losers in the assessment of climate change impacts. But those implications are not simple. The climate system is inter- and teleconnected. Therefore, to some degree the issue can be dealt with as a local phenomenon laid over a global phenomenon; but sulfur also has global commons properties in the same way as CO_2. The construction of a complete climate-change story may be leading inexorably to the construction of a global-change story. To adequately address welfare economic issues arising from the climate-change problem may mean jointly addressing acid deposition and climate change. Clearly, the sulfur issue does not admit to easy dismissal. As a policy issue it derives from the work of IA modelers and, while complex and difficult, will likely remain a matter for their attention.

BIGGER, FASTER, SMALLER, BETTER

Kolstad's final consideration is the future direction of IA modeling. The activity of forecasting the future of research is an indulgence in which we, who are IA modelers, view the world through the tinted lenses of our own research agendas. I expect the pattern of IA research to reflect the product of interactions between the policy process and the IA modeling community in which the former generates questions, IAMs generate results, which in turn influence and alter the next set of questions emerging from the policy process. As nations, organizations, and individuals prepare for the 1997 Conference of the Parties in Japan, there will doubtless be a wide suite of potential agreements upon which IA modelers will be asked to comment. These comments in turn will play a role in sorting through alternatives and the reformulation of proposals. While IAMs will have a part in the process, they will have a supporting rather than leading role. It is sobering to observe that no IA modelers, and virtually no scientists of any stripe, participated in the Rio negotiations of the original FCCC.

But to have any role at all, IAMs must evolve to consider issues that are relevant. At this time the research trail leads in many directions, providing opportunities for a variety of IAM approaches to contribute. Most of the trails involve increasingly sophisticated IAMs. The larger, process-oriented IAMs are beginning to produce results, and their experience will enable the more highly aggregated models to develop more sophisticated representations. In the years ahead IAMs will gradually move from a framework in which climate-change damages are treated as a stylized variable to a framework in which component damages are addressed individually. There will be an increasingly sophisticated treatment of a broader suite of greenhouse-related emissions. There will be increasingly sophisticated treatments of the natural-social-science interface, as for example in the areas of agriculture, land-use change, and energy. Models will improve their characterization of decisionmaking under uncertainty. And there will be some surprises.

CONCLUDING REMARKS

IA is in the midst of an extraordinary period of intellectual growth and excitement. Over a very short time it has exerted an extraordinary influence on the formulation of the climate issue. This sense of excitement is reflected in both Weyant and others (1996) and Kolstad's review. Kolstad's main conclusion that "[c]onsidering the relatively short history of integrated assessment, ... a surprising amount of knowledge has emerged" is one to which I heartily subscribe.

REFERENCES

Alcamo, J., ed. 1994. *IMAGE 2.0: Integrated Modeling of Global Climate Change*. Dordrecht: Kluwer Academic Publishers.

Alcamo, J., R. Shaw, and L. Hordijk, eds. 1990. *The RAINS Model of Acidification: Science and Strategies in Europe*. Dordrecht: Kluwer Academic Publishers.

Ball, M. and H. Dowlatabadi. 1995. The Role of Aerosols in Climate Change: Results from an Integrated Assessment Model (ICAM-2.0). Submitted to *Environmental Policy*, Carnegie Mellon University, Department of Engineering and Public Policy.

Dowlatabadi, H. and M. G. Morgan. 1993. A Model Framework for Integrated Studies of the Climate Problem. *Energy Policy* 21:209–21.

Edmonds, J., H. Pitcher, N. Rosenberg, and T. Wigley. 1993. *Design for the Global Change Assessment Model*. Presented to the *International Workshop on Integrative Assessment of Mitigation, Impacts and Adaptation to Climate Change*, October 13–15, Laxenburg, Austria.

Edmonds, J. and J. Reilly. 1985. *Global Energy: Assessing the Future*. New York: Oxford University Press, 317.

Edmonds J. and M. Wise. 1996. Stabilizing Atmospheric CO_2: Rethinking the Emissions Problem. In *An Economic Perspective on Global Climate Change*. Washington, D.C.: American Council for Capital Formation.

Edmonds, J., M. Wise, and C. MacCracken. 1994. *Advanced Energy Technologies and Climate Change: An Analysis Using the Global Change Assessment Model (GCAM)*. Proceedings of the Air and Waste Management Meeting on climate change, April 6, Tempe, Arizona.

Edmonds, J., M. Wise, H. Pitcher, R. Richels, T. Wigley, and C. MacCracken. 1996. An Integrated Assessment of Climate Change and the Accelerated Introduction of Advanced Energy Technologies: An Application of MiniCAM 1.0. *Mitigation and Adaptation Strategies for Global Change* 1(4):311–39.

IPCC (Intergovernmental Panel on Climate Change). 1992. *Climate Change 1992: The Supplementary Report to the IPCC Scientific Assessment*, edited by J. T. Houghton, B. A. Callander, and S. K. Varney. Cambridge: Cambridge University Press.

———. 1996a. *Climate Change 1995: The Science of Climate Change. The Contribution of Working Group I to the Second Assessment Report of the Intergovernmental Panel on Climate Change*, edited by J. P. Houghton, L. G. Meira Filho, B. A. Callendar, A. Kattenberg, and K. Maskell. Cambridge: Cambridge University Press.

———. 1996b. *Climate Change 1995: Impacts, Adaptation, and Mitigation of Climate Change: Scientific-Technical Analysis. The Contribution of Working Group II to the Second Assessment Report of the Intergovernmental Panel on Climate Change*, edited by R. T. Watson, M. C. Zinyowera, and R. H. Moss. Cambridge: Cambridge University Press.

———. 1996c. *Climate Change 1995: Economic and Social Dimensions of Climate Change. The Contribution of Working Group III to the Second Assessment Report of the Intergovernmental Panel on Climate Change*, edited by J. P. Bruce, H. Lee, and E. F. Haites. Cambridge: Cambridge University Press.

Kelly, D. L. and C. D. Kolstad. 1996. *The Climate Change Footprint: Will We See It Before It Is Upon Us?* Department of Economics, University of California–Santa Barbara.

Kolstad, C. D. 1992. Looking vs. Leaping: The Timing of CO_2 Control in the Face of Uncertainty and Learning. In *Costs, Impacts, and Possible Benefits of CO_2 Mitigation*, edited by Y. Kaya, N. Nakicenovic, W. D. Nordhaus, and F. L. Toth. The Institute for Applied Systems Analysis (IIASA), Laxenburg, Austria (June).

———. 1994. The Timing of CO_2 Control in the Face of Uncertainty and Learning. In *International Environmental Economics*, edited by E. C. Van Ierland. Amsterdam: Elsevier.

———. 1996. *Déjà Vu All Over Again: What's New and What's Not in Integrated Assessment Modeling of Climate Change*. Presented to the National Bureau of Economic Research, Snowmass, Col.

Manne, A. S., R. Mendelsohn, and R. Richels. 1995. MERGE—A Model for Evaluating Regional and Global Effects of GHG Reduction Policies. *Energy Policy* 23(1):17–34.

Morita, T., Y. Matsuoka, M. Kainuma, H. Harasawa, K. Kai, and S. Nishioka. 1993. *AIM: Asian-Pacific Integrated Model for Evaluating Policy Options to Reduce GHG*

Emissions and Global Warming Impacts. Tsukuba, Ibaraki, Japan: National Institute for Environmental Studies.

Nordhaus, W. D. 1979. *The Efficient Use of Energy Resources.* New Haven, Connecticut: Yale University Press.

———. 1994. *Managing the Global Commons: The Economics of Climate Change.* Cambridge, Massachusetts: MIT Press.

Prinn, R., H. Jacoby, A. Sokolov, C. Wang, X. Xiao, Z. Yang, R. Eckaus, P. Stone, D. Ellermand, J. Melillo, J. Fitzmaurice, D. Kicklighter, Y. Liu, and G. Holian. 1996. *Integrated Global System Model for Climate Policy Analysis: I. Model Framework and Sensitivity Studies.* Report No. 7. MIT Joint Program on the Science and Policy of Global Change, Cambridge, Massachusetts.

Richels, R., J. Edmonds, H. Gruenspecht, and T. Wigley. 1996. *The Berlin Mandate: The Design of Cost-Effective Mitigation Strategies.* Report of the Energy Modeling Forum Subgroup on Regional Distribution of the Costs and Benefits of Climate Change Policy Proposals. Stanford, California.

Rotmans, J. 1990. *IMAGE: An Integrated Model to Assess the Greenhouse Effect.* Dordrecht: Kluwer Academic Publishers.

Wang, W.-C., D. J. Wuebbles, and W. M. Washington. 1985. Potential Climatic Effects of Perturbations Other Than Carbon Dioxide. In *Projecting the Climate Effects of Increasing Carbon Dioxide,* edited by M. C. MacCracken and F. M. Luther. DOE/ER-0237. Springfield, Virginia: National Technical Information Service, U.S. Department of Commerce.

Weyant, J., O. Davidson, H. Dowlatabadi, J. Edmonds, M. Grubb, E. A. Parson, R. Richels, J. Rotmans, P. R. Shukla, and R. S. J. Tol. 1996. Integrated Assessment of Climate Change: An Overview and Comparison of Approaches and Results. Chapter 10 in IPCC 1996c.

Wigley, T. M. L., R. Richels, and J. A. Edmonds. 1996. Economic and Environmental Choices in the Stabilization of Atmospheric CO_2 Concentrations. *Nature* 379(6562):240–43.

Wise, M. and J. Edmonds. 1996. *Efficient Strategies for OECD CO_2 Emissions Reductions Targets.* Proceedings of the NATO conference on Sustainable Development, Durham, North Carolina.

Index

Milton Keynes UK
Ingram Content Group UK Ltd.
UKHW031143141024
449569UK00024B/1104